D1423358

Dedication

This book is dedicated to Sir Ronald Nyholm, F.R.S., who died in a tragic accident on 4th December, 1971. Nyholm's own pioneering investigations into the preparation and structures of new types of transition-metal complexes and his enormous enthusiasm for inorganic chemistry inspired many others to follow in his footsteps and made him one of the leaders of the renaissance of inorganic chemistry that commenced in the 1950's.

The author was privileged to have been a friend and colleague of Sir Ronald Nyholm at University College, London, during this important and exciting time. Their many discussions on the structures of inorganic molecules led to the publication in 1957 of a joint paper entitled *Inorganic Stereochemistry*. The ideas in that paper have been subsequently modified and expanded but nevertheless form the foundation on which the present book is based.

It seems fitting therefore to dedicate this book as a small tribute to an inspiring leader in inorganic chemistry whose tragic loss will be felt for many years to come. His contributions to, and his influence upon, the subject of inorganic chemistry will live on.

MOLECULAR GEOMETRY

R. J. GILLESPIE

Professor of Chemistry
McMaster University
Hamilton, Ontario

VAN NOSTRAND REINHOLD COMPANY
LONDON
NEW YORK CINCINNATI TORONTO MELBOURNE

VAN NOSTRAND REINHOLD COMPANY LTD
Windsor House, 46 Victoria Street, London SW1

INTERNATIONAL OFFICES
New York Cincinnati Toronto Melbourne

Library of Congress Catalog Card No: 77–160198

First published 1972

PRINTED IN GREAT BRITAIN BY
BUTLER & TANNER LTD.
FROME AND LONDON

Preface

Stereochemistry or three-dimensional structural chemistry was born in 1874 with van't Hoff and Le Bel's postulate of the tetrahedral carbon atom. Since that time the development of various spectroscopic and diffraction methods of studying molecules and crystals has provided a wealth of information on their structures, that is, on the arrangement of their atoms in space.

It has been found that many other atoms also have a tetrahedral arrangement of four bonds in their compounds, but that four bonds sometimes adopt a square planar arrangement, that three bonds often have a pyramidal arrangement, but sometimes a planar arrangement, and that six bonds generally adopt an octahedral arrangement. Theory has not, however, kept up with experiment, and no comprehensive and completely satisfactory theory for understanding and predicting the structures of molecules has been developed. This is not to imply that we have no understanding of chemical bonding; indeed, several detailed and impressive theories have been developed, but it is none the less true to say that they have not been completely successful in providing a basis for understanding and predicting why one particular stereochemistry is preferred to another. The prediction by several theoretical chemists that XeF_6 would have an octahedral structure—which has been found not to be the case—is but one example of many that could be cited of the inadequacy of theory in making predictions of this kind.

The purpose of this book is to develop a theory, or more exactly a set of rules, for predicting molecular geometry based on the idea that the arrangement in space of the covalent bonds formed by an atom depends primarily on the arrangement of the electron pairs in the valency shell of the atom which is determined mainly by the operation of the Pauli Exclusion principle. Although these rules are admittedly somewhat empirical, they do have a quantum mechanical basis, namely the Pauli Exclusion principle and, in any case, an important justification for them is that they provide a simple and reliable basis for understanding and predicting molecular geometry. It is of course not possible to discuss more than a small fraction of the very large number of molecular structures that has been determined in recent years. However, a representative selection has been made of all the most important structural types.

A number of the basic ideas of the theory presented in the book were developed from some suggestions first made by N. V. Sidgwick and H. E. Powell in an important paper published in 1940 (*Proc. Roy. Soc.*, **A 176,** 153 (1940)) and which were first formulated in vigorous and lively discussions with Professor Sir Ronald Nyholm, F.R.S., some fifteen years ago and published in a joint paper in 1957 (*Quart. Rev. Chem. Soc.*, **11,** 339 (1957)).

Since that time the ideas have gained some acceptance and have become known as the Valence Shell Electron Pair Repulsion Theory. Elementary accounts of the theory together with a discussion of a limited number of examples have appeared in a number of general chemistry and inorganic chemistry textbooks and the author has discussed a number of special developments of the theory in journal articles. However no general account of the theory or comprehensive discussion of its applications throughout the Periodic Table have been given since the original 1957 paper. It is the purpose of this book to give an up-to-date and comprehensive account of the theory and to discuss a wide variety of applications.

I would like to express my indebtedness to the original work of Sidgwick and Powell and also my gratitude to Sir Ronald Nyholm for inspiring my interest in the fascinating and varied structures of inorganic molecules.

I would also like to express my thanks to three other people without whose help this book would not have been written: my wife for her continued encouragement and for making sure that I had the necessary time, Miss Peggy McLauchlin for her flawless typing and for her ability to decipher my handwriting, and Mrs. Gay Parsons Walker for transforming my illegible sketches into finished diagrams.

Contents

1

The Chemical Bond

1.1 ELECTRONIC STRUCTURES OF ATOMS

It is known from ionization energies and from the form of the periodic table of the elements that the electrons in an atom are arranged in shells. With increasing distance from the nucleus successive shells have increasing energies; are of increasing size; and contain larger numbers of electrons. The innermost K shell contains a maximum of only two electrons, the next shell, the L shell, a maximum of

Table 1.1

Elements	Atomic number or nuclear charge (Z)	Electron shells K	L	M
H	1	1		
He	2	2		
Li	3	2	1	
Be	4	2	2	
B	5	2	3	
C	6	2	4	
N	7	2	5	
O	8	2	6	
F	9	2	7	
Ne	10	2	8	
Na	11	2	8	1
Mg	12	2	8	2
Al	13	2	8	3
Si	14	2	8	4
P	15	2	8	5
S	16	2	8	6
Cl	17	2	8	7
Ar	18	2	8	8

eight electrons, and the M shell a maximum of eighteen electrons. The number of electrons is in fact given by $2n^2$ where n is a quantum number which is 1 for the K shell, 2 for the L shell and so on. The electronic configurations for the first few elements may therefore be written as in Table 1.1.

1.2 THE ELECTRON-PAIR BOND: LEWIS DIAGRAMS

Following the suggestion of G. N. Lewis, the chemical bond or covalent bond became identified with a shared pair of electrons, and it has since become clear that it is the electrostatic attraction of this pair of electrons for the two nuclei that holds the two nuclei together in the chemical bond. The completed inner shells of electrons, together with the nucleus, constitute a spherical inner core of the atom that is not, in general, involved in bonding. Thus, for carbon, the nucleus of charge +6 plus the two L-shell electrons, constitute the inner core which has a resultant charge of +4. The core charge clearly increases from +1 for lithium to +8 for neon. For sodium, the core charge drops to +1 because the inner core now consists of the nucleus of charge +11 plus the K and L shells which contain a total of ten electrons. The valencies of the elements Li to Ne can be understood if it is assumed that the valence shell is subdivided into four regions,

	Li	Be	B	C	N	O	F	Ne
Core charge	+1	+2	+3	+4	+5	+6	+7	+8
Number of electrons in valence shell	1	2	3	4	5	6	7	8
Arrangement of electrons in valence shell surrounding positive inner core	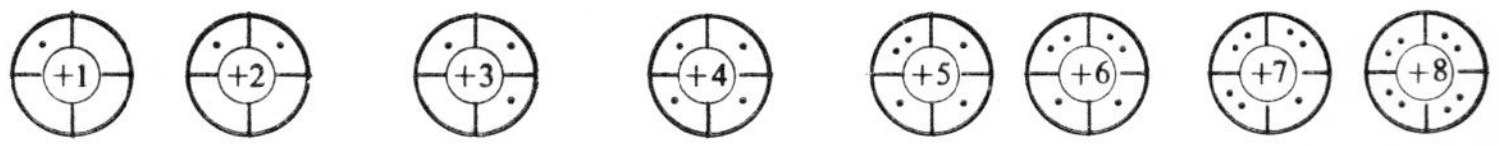 (+1)	(+2)	(+3)	(+4)	(+5)	(+6)	(+7)	(+8)
Valence or number of bonds = n or 8 − n	1	2	3	4	3	2	1	0
Electron dot representation	Li·	·Be·	·B̈·	·Ċ̣·	·N̈̇·	:Ö·	:F̤̈·	:N̤̈e:
Lewis diagrams for the fluorides	Li:F:	:F:Be:F:	:F:B:F: (F above B)	:F:C:F: (F above and below C)	:F:N:F: (F above N)	:O:F: (F below O)	:F:F:	

usually called orbitals, which can each accommodate two electrons, and if each orbital is singly filled before it accommodates a second electron. The arrangements of the electrons in the valence shells of the elements Li to Ne can then be illustrated in the manner on page 2.

The positive charge of the inner core can attract additional electrons into the valence shell if it is incomplete, i.e., if there are one or more singly filled orbitals. Thus the oxygen atom may acquire two additional electrons to give the O^{2-} ion, and the fluorine atom may acquire one additional electron to give the fluoride ion F^- so that in each case the L shell is completed.

Alternatively if these additional electrons are acquired from one or more other atoms transfer of the electrons may not be complete and an atom may share one or more pairs of electrons with other atoms, e.g.

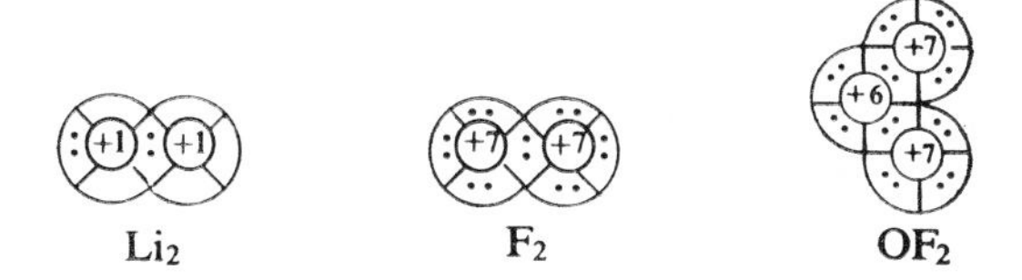

Each pair of shared electrons constitutes a covalent bond and an atom in general forms as many bonds as it has singly occupied orbitals. In this way the elements carbon, oxygen, and fluorine each acquire filled valence shells of eight electrons as found in the noble gases neon and argon which have no singly occupied orbitals and are not known to form any compounds. Thus an atom with n electrons in its valence shell forms either n or $8 - n$ covalent bonds. The tendency of an atom to acquire a stable outer shell of eight electrons is called the octet rule. These ideas concerning the covalent bond and the valencies of the elements were first clearly formulated by G. N. Lewis, and the arrangements of the electrons in molecules may be clearly represented by electron-dot diagrams or Lewis diagrams in which the electrons are represented by dots arranged singly or in pairs according as to whether they occupy singly filled or doubly filled orbitals. The electron pairs may be distinguished as bonding pairs which are shared between two nuclei and non-bonding pairs or lone-pairs which are located only on one nucleus. It is the electrostatic force of attraction of a pair of electrons situated between two nuclei for these two nuclei that holds the two nuclei together in the covalent bond.

The elements lithium, beryllium, and boron have insufficient electrons to complete the octet in neutral molecules, even when they are all used for bond formation, although they can complete their valence shells by forming ions such as BF_4^- and BeF_4^{2-}.

```
[    ··     ]-            [     ··      ]2-
[   :F:     ]             [    :F:      ]
[ ··  ··    ]             [ ··  ··   ·· ]
[:F : B : F:]             [:F : Be : F: ]
[ ··  ··    ]             [ ··  ··   ·· ]
[   :F:     ]             [    :F:      ]
[    ··     ]             [     ··      ]
```

In fact, the octet rule applies without exception *only* to the elements carbon, nitrogen, oxygen, and fluorine, for which the valence shell is the L shell, which can only accommodate a maximum of eight electrons. Hydrogen is normally restricted to the formation of the one bond needed to complete its K shell of two electrons, the elements lithium, beryllium, and boron often do not complete their valence shells in molecule formation, and the heavier elements from sodium on may, and often do, have more than eight electrons in their valence shells. Thus elements such as phosphorus and sulphur, for which the valence shell is the M shell, and which can contain up to eighteen electrons, can use all their electrons in bond formation, thus achieving valence shells containing ten or twelve electrons as in PF_5 or SF_6.

Although the octet rule has many exceptions, it nevertheless played a very important role in the development of our understanding of chemical bonding. It is clear that if an element forms bonds by sharing pairs of electrons to, as far as possible, complete its valence shell, then all compounds should contain an even number of electrons, and it is indeed true that the vast majority of stable molecules do contain an even number of electrons. The electrons in the valence shell of an atom in a compound may then be conveniently regarded as being arranged in pairs, some of which are forming chemical bonds, and are described as bonding pairs, and others of which are not forming bonds and may be described as non-bonding pairs or lone-pairs. The basis of the discussion of molecular geometry in this book is that the stereochemistry of an atom, i.e., the arrangement of its covalent bonds in space, depends in the first instance only on the numbers of bonding and non-bonding electron pairs in its valence shell.

1.3 IONIC AND COVALENT BONDS: ELECTRONEGATIVITY

This book is largely confined to a discussion of simple molecules of the type AX_n in which n atoms or groups of atoms, called ligands, are bound to a central atom A by covalent or predominately covalent bonds. More complex molecules and infinite lattices are discussed only briefly, as no new principles are involved. Because different atoms, by virtue of their different core charges and different sizes, attract electrons more or less strongly, covalent bonds between different atoms are, in general, polar, i.e., they have a certain amount of ionic character because the bonding pair of electrons is not shared equally between the two atoms but is located somewhat closer to the atom that has the stronger attraction for electrons in its valence shell. The power of an atom to attract electrons in its valence shell is called its electronegativity. The concept of electronegativity cannot be rigorously defined and a variety of methods have been proposed for obtaining values for the electronegativities of the elements. The most widely used values are those due to Pauling which are given in Table 1.2. Values given by other authors in some cases differ slightly from these values. Moreover the electronegativity of an element varies somewhat with the nature of the attached ligands and with the oxidation state of the element. We will in any case only use these values to obtain a qualitative idea of the polarity of a given bond.

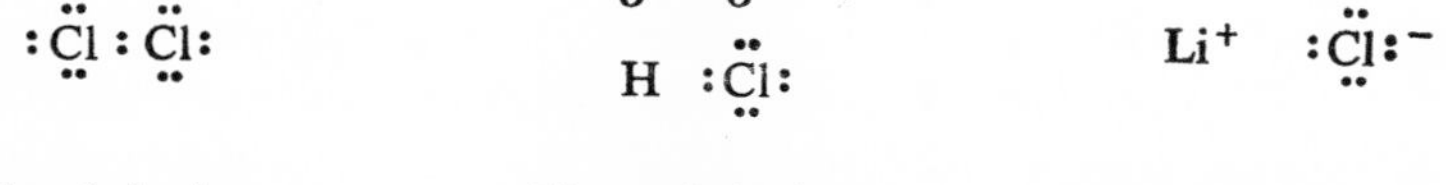

:Cl:Cl:	δ^+ H δ^- :Cl:	Li^+ :Cl:$^-$
Equal sharing of bonding electron pairs (Non-polar bond).	Unequal sharing of bonding electron pair (Polar bond).	Very unequal sharing of bonding electron pair ('Ionic' bond).

Thus the bond in Cl_2 is a non-polar covalent bond, while the bond in HCl is a polar covalent bond because chlorine attracts the electrons of the bond more strongly than hydrogen. If the electronegativities of the two bonded atoms are very different, as is the case for lithium and chlorine, then the bond is very polar and is usually described as an ionic bond in which a positive ion (e.g., Li^+) is held by electrostatic attraction to the negative Cl^- ion. Such highly polar molecules can only exist independently at low concentrations in the gas phase because they attract each other very strongly and they combine together to form an infinite solid lattice, the structure of which is best understood in terms of the packing of charged spheres, being determined primarily by the relative sizes and the charges of the ions. Such ionic lattices are not specifically considered in this book as the ideas presented here are not directly relevant to such structures. However, it must be remembered that a pure ionic bond is strictly a limiting case and most bonds have more or less covalent character. Indeed in many so-called ionic lattices there is considerable covalent character and the principles discussed in this book can indeed be applied to these structures.

1.4 THE ARRANGEMENT OF ELECTRON PAIRS IN VALENCE SHELLS

It was first suggested by Sidgwick and Powell in 1940 that molecular geometry was determined by the arrangement of electron pairs in the valence shell, and this suggestion has subsequently been developed into a set of rules known as the *valence-shell electron-pair repulsion* theory, which enable many features of molecular structure to be predicted and understood in a simple manner. The first and most fundamental rule can be stated as follows:

The pairs of electrons in a valence shell adopt that arrangement which maximizes their distance apart, i.e., the electron pairs behave as if they repel each other.

We assume for the present that the inner shells are complete, and therefore the central core of the atom consisting of the nucleus and the completed inner shells is spherical and has no effect on the distribution of the outer, or valence, electrons. The consequences of a non-spherical central core, as is often found for the transition elements, are discussed later. Making the further simplifying assumption that the electron pairs in a valence shell are all at the same

Table 1.2

Electronegativity values of the elements

							H									
							2·1									
Li	Be											B	C	N	O	F
1·0	1·5											2·0	2·5	3·0	3·5	4·0
Na	Mg											Al	Si	P	S	Cl
0·9	1·2											1·5	1·8	2·1	2·5	3·0
K	Ca	Sc	Ti	V	Cr	Mn	Fe	Co	Ni	Cu	Zn	Ga	Ge	As	Se	Br
0·8	1·0	1·3	1·5	1·6	1·6	1·5	1·8	1·8	1·8	1·9	1·6	1·6	1·8	2·2	2·4	2·8
Rb	Sr	Y	Zr	Nb	Mo	Tc	Ru	Rh	Pd	Ag	Cd	In	Sn	Sb	Te	I
0·8	1·0	1·2	1·4	1·6	1·8	1·9	2·2	2·2	2·2	1·9	1·7	1·7	1·8	1·9	2·1	2·5
Cs	Ba	La–Lu	Hf	Ta	W	Re	Os	Ir	Pt	Au	Hg	Tl	Pb	Bi	Po	At
0·7	0·9	1·1–1·2	1·3	1·5	1·7	1·9	2·2	2·2	2·2	2·4	1·9	1·8	1·8	1·9	2·0	2·2
Fr	Ra	Ac	Th	Pa	U	Np–No										
0·7	0·9	1·1	1·3	1·5	1·7	1·3										

Source: Linus Pauling, *Nature of the Chemical Bond*, 3rd ed., Cornell University Press (1960).

average distance from the nucleus, the arrangements for two to six electron pairs which maximize their distance apart are as shown in Table 1.3 and Fig. 1.1. If each electron pair is represented by a point

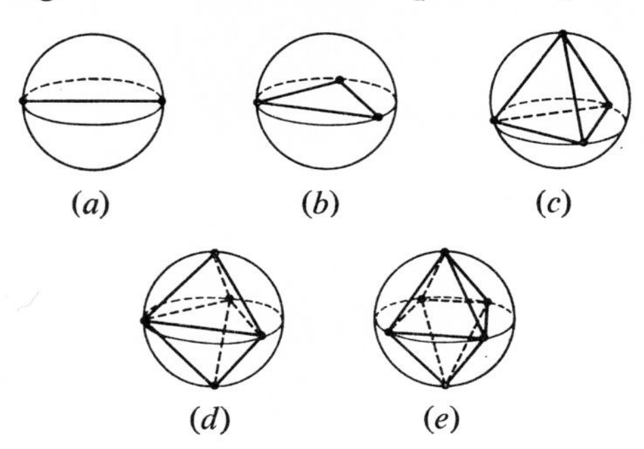

FIG. 1.1 Arrangements of points on the surface of a sphere that maximize their distance apart: (*a*) linear arrangement of two points; (*b*) equilateral triangular arrangement of three points; (*c*) tetrahedral arrangement of four points; (*d*) trigonal bipyramidal arrangement of five points; (*e*) octahedral arrangement of six points.

then the arrangements are described by the polyhedra formed by connecting the points by straight lines.

The tetrahedron and octahedron are regular polyhedra, but the trigonal bipyramid is not: it has five vertices, nine edges, and six

Table 1.3

Number of electron pairs	Arrangement
2	linear
3	equilateral triangle
4	tetrahedron
5	trigonal bipyramid
6	octahedron

triangular faces, but the vertices and edges are not all equivalent. (Fig. 1.2.)

The arrangements for two, three, four, and six electron pairs are intuitively obvious, but they can be obtained in a rigorous manner by considering the arrangement of a given number of points on the surface of a sphere (each point representing one electron pair) which maximizes the least distance between any pair of points. The solutions to this problem are as given above, except that for five points the solution is indeterminate, as both the trigonal bipyramid and the square pyramid and any intermediate arrangement are possible solutions. However, if we also minimize the number of such least

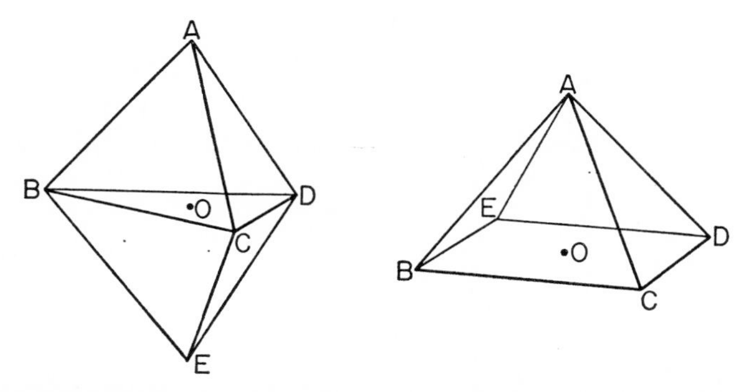

FIG. 1.2 The trigonal bipyramid and the square pyramid. O is the centre of the circumscribing sphere of radius r. For the trigonal bipyramid AB = AC = AD = BE = CE = DE = $\sqrt{2}r$ and BD = DC = CB = $\sqrt{3}r$. For the square pyramid all eight edges = $\sqrt{2}r$.

distances, then the trigonal bipyramid, which has six $\sqrt{2}r$ interparticle distances, where r is the radius of the sphere, is favoured over the square pyramid which has eight such distances.

1.5 SHAPES OF MOLECULES

Each of the arrangements of a given number of electron pairs in Table 1.3 can give rise to several molecular shapes, depending on the number of bonding and non-bonding electron pairs. If we let the central atom be A, X a ligand and E a non-bonding pair or lone-pair then in a singly bonded molecule AX_mE_n there are $m + n$ electron pairs in the valence shell, of which m are bonding pairs and n are non-bonding or lone-pairs. The shape of the molecule is determined by the most probable arrangement of $m + n$ electron pairs.

Two electron pairs have a linear arrangement. If both electron pairs are bonding pairs a linear molecule results, e.g., beryllium dichloride in the gas phase (1).

Cl—Be—Cl

(1)

Three electron pairs have an equilateral triangular arrangement.

An AX_3 molecule in which all three electron pairs are bonding has a planar triangular shape, as has boron trifluoride (2).

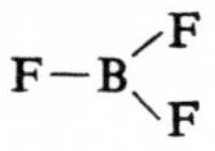

```
     /F
F—B
     \F
```

(2)

An AX_2E molecule has an angular shape—stannous chloride in the gas phase (3) is an example.

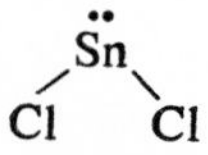

```
   ••
   Sn
  /  \
Cl    Cl
```

(3)

Four electron pairs have a tetrahedral arrangement. Therefore an AX_4 molecule is tetrahedral, e.g., methane (4).

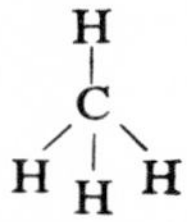

```
   H
   |
   C
  /|\
 H H H
```

(4)

An AX_3E molecule has a pyramidal shape with a non-bonding electron pair occupying one of the tetrahedral positions as in the ammonia molecule (5).

```
 ••
 N
/|\
HHH
```

(5)

An AX_2E_2 molecule has an angular shape with two non-bonding electron pairs occupying corners of the tetrahedron as in the water molecule (6).

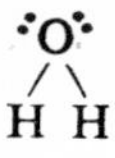

```
:O:
/ \
H H
```

(6)

Five electron pairs have a trigonal bipyramid arrangement, hence an AX_5 molecule has a trigonal bipyramid shape, e.g., PCl_5

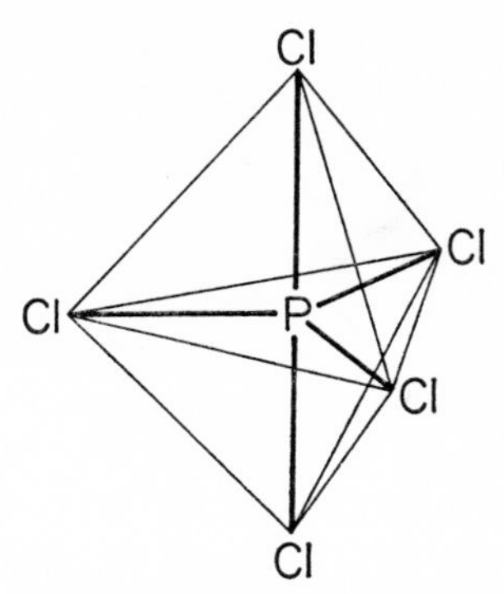

FIG. 1.3 The trigonal bipyramidal PCl_5 molecule.

(Fig. 1.3). The trigonal bipyramid is the first case we meet of a polyhedron with vertices that are not all equivalent. The two axial vertices (*a* in Fig. 1.4) are not equivalent to the three equatorial vertices (*e*

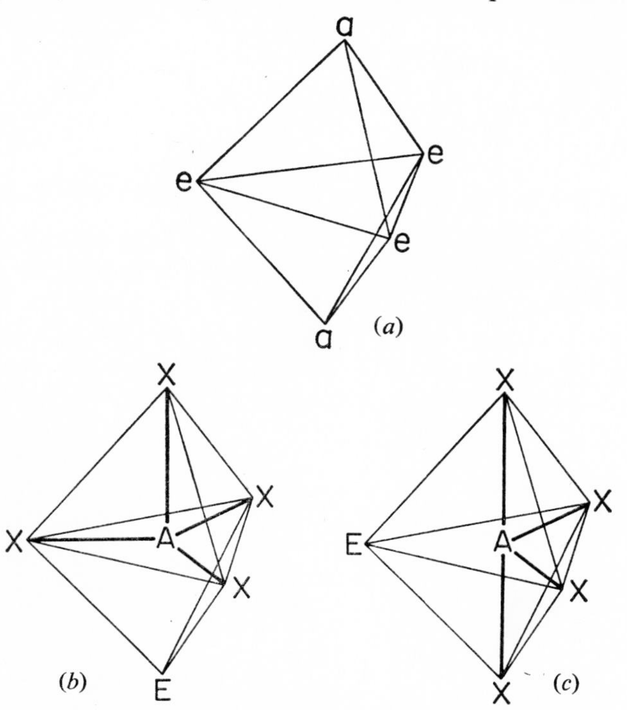

FIG. 1.4 (*a*) The axial 'a' and equatorial 'e' positions of a trigonal bipyramid; (*b*) and (*c*) are alternative shapes for an AX_4E molecule. The preferred shape is (*c*).

in Fig. 1.4) as the axial vertices each have three adjacent vertices at 90°, i.e., the three vertices *e*, while each equatorial vertex has two adjacent vertices at 90°, and two adjacent vertices at 120°. This has a number of important consequences which are discussed in greater detail in Chapter 4. Of interest to us now is the fact that there are alternative non-equivalent positions for a lone-pair in an AX_4E molecule, i.e., an axial or an equatorial position. For reasons that are discussed later (Chapter 3) lone-pairs always occupy the equatorial positions and therefore an AX_4E molecule has the shape *c* rather than *b* in Fig. 1.4. Sulphur tetrafluoride is an example of a molecule having this shape, which is that of an irregular tetrahedron or

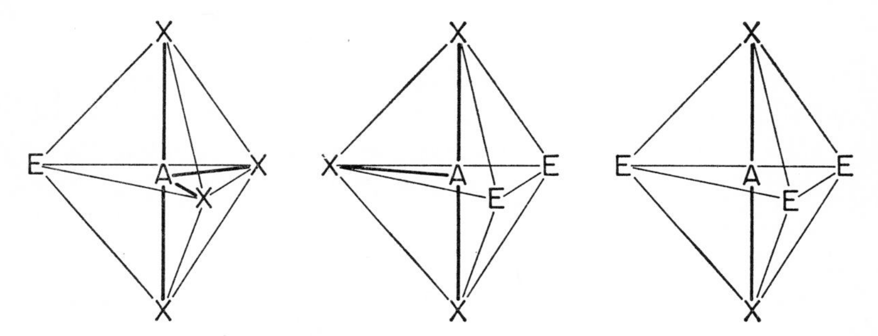

FIG. 1.5 Shapes of AX_4E, AX_3E_2 and AX_2E_3 molecules.

disphenoid. In an AX_3E_2 molecule both lone-pairs occupy equatorial positions, and therefore such molecules have the shape shown in Fig. 1.5, in which all three ligands X are in the same plane but where they make two angles of 90° and one angle of 180° with each other instead of three angles of 120° as in an AX_3 molecule. An example of such a molecule is ClF_3. Finally, an AX_2E_3 molecule such as

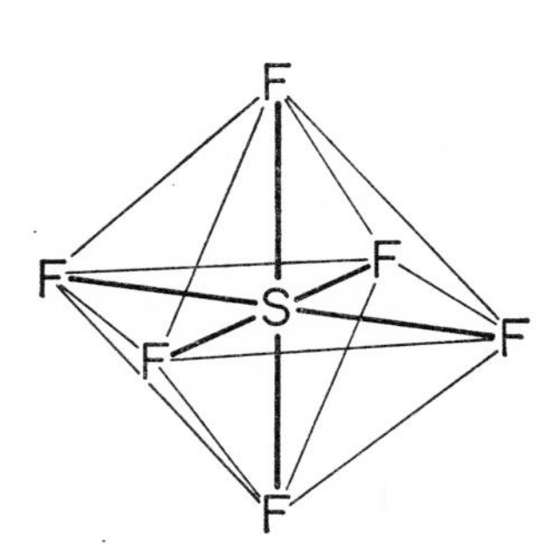

FIG. 1.6 The octahedral SF_6 molecule.

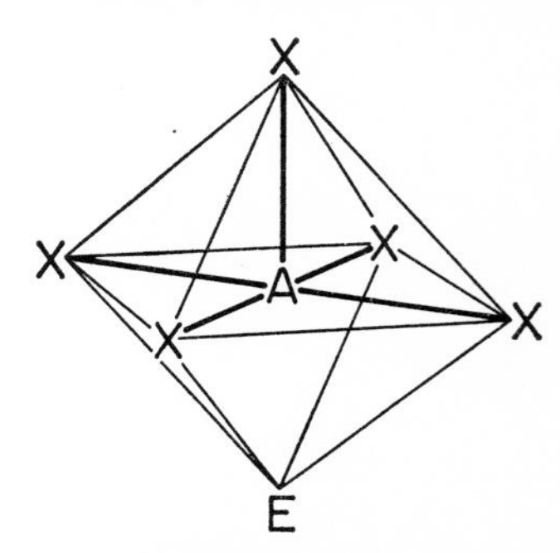

FIG. 1.7 Square pyramidal AX_5E molecule.

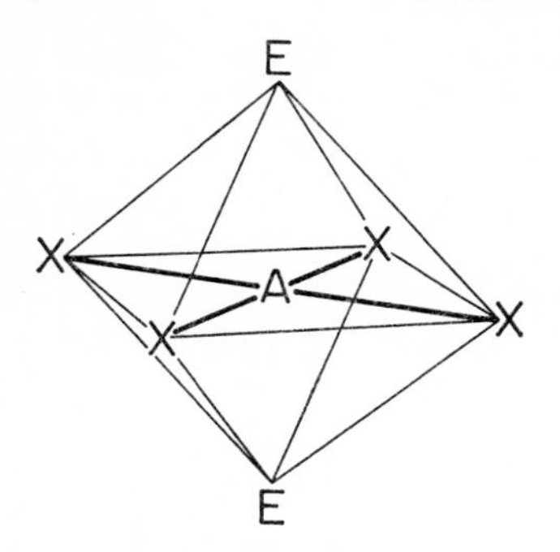

FIG. 1.8 Square planar AX_4E_2 molecule.

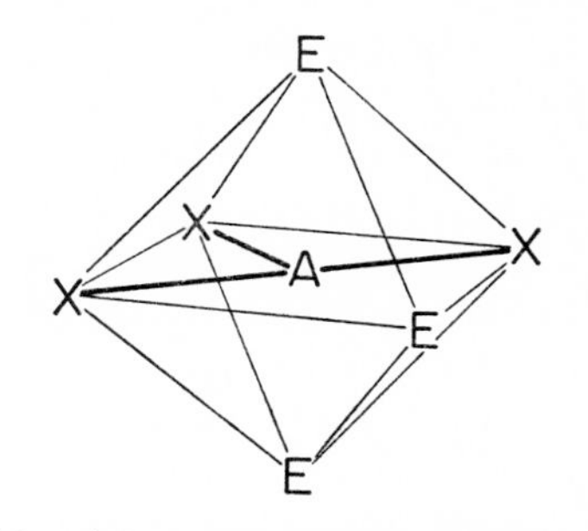

FIG. 1.9 Predicted shape for an AX_3E_3 molecule.

XeF_2 has a linear shape arising from bonding electron pairs in the axial positions and all three lone-pairs in the equatorial positions.

AX_6 molecules are octahedral, e.g., sulphur hexafluoride (Fig. 1.6). AX_5E molecules have a square pyramid shape with the lone-pair occupying the sixth octahedral position as in IF_5 (Fig. 1.7).

Table 1.4

Number of electron pairs	Arrangement	Number of lone-pairs	Type of molecule	Shape of molecule	Examples
2	Linear	0	AX_2	Linear	$BeCl_2$
3	Equilateral triangle	0	AX_3	Equilateral triangle	BF_3
		1	AX_2E	V-shape	$SnCl_2$
4	Tetrahedron	0	AX_4	Tetrahedron	CCl_4
		1	AX_3E	Trigonal pyramid	NF_3
		2	AX_2E_2	V-shape	H_2O
5	Trigonal bipyramid	0	AX_5	Trigonal bipyramid	PCl_5
		1	AX_4E	SF_4-shape*	SF_4
		2	AX_3E_2	T-shape	ClF_3
		3	AX_2E_3	Linear	XeF_2
6	Octahedron	0	AX_6	Octahedron	SF_6
		1	AX_5E	Square pyramid	IF_5
		2	AX_4E_2	Square	XeF_4

* This asymmetrical shape has no convenient name. The ideal shape shown in Fig. 1.10 is a bisphenoid (C_{2v} symmetry), but as discussed in Chapter 4 the actual molecule is slightly distorted from this shape. Thus it is most conveniently referred to as the SF_4-shape.

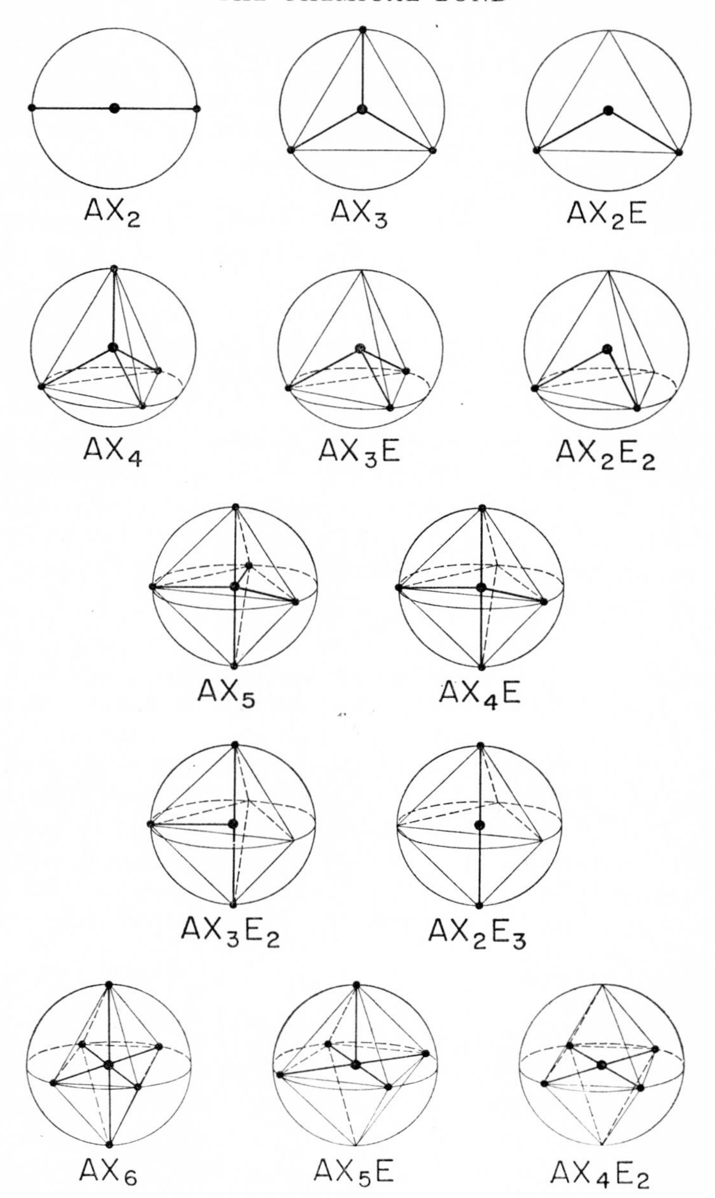

FIG. 1.10 Predicted shapes for all molecules containing up to six electron pairs in their valence shells and having a spherical inner shell.

In AX_4E_2 molecules there are alternative positions for the lone-pairs E; they may be either *cis* or *trans* to each other. For reasons discussed later they are always *trans*: giving rise to a square planar molecule, e.g., XeF_4 (Fig. 1.8). There are no known AX_3E_3 molecules, but they might be predicted to have the structure shown in Fig. 1.9.

The general shapes of all singly-bonded molecules based on valence shells containing two to six electron pairs and a spherical inner shell are summarized in Table 1.4 and Fig. 1.10.

According to the theory presented in this chapter, the geometry of a molecule, that is to say the arrangement of ligands around a central atom, is determined by the preferred arrangement of the electron pairs in the valence shell of the central atom. Possible interactions between the ligands have been ignored, and it is assumed that these are nearly always of lesser importance than the valence shell electron-pair repulsions.

1.6 REGULAR POLYHEDRA

As we have just seen, the arrangement of the bonds around a central atom is most conveniently described by connecting the ends of the bonds to form a polyhedron. The first such polyhedron that was recognized to be of importance in chemistry was the tetrahedron, which is a regular solid with four equilateral triangular faces, four vertices, and four edges. The regular solids (or polyhedra) may be defined as those polyhedra that have all corner angles equivalent to one another, all edges equivalent, and in which all the faces are regular polygons and are equivalent. There are five such regular solids. They were fully described by the Greeks and were probably known much earlier—a pentagonal dodecahedron having been discovered in an Etruscan tomb.

It is easy to demonstrate that there are only five regular solids, and to show what their shapes must be. In order to have a solid angle, at least three edges must meet at a point. If three equal-length edges meet at every apex, and if they join each other in triangular faces, the solid obtained is a tetrahedron, which has four faces and six edges (Fig. 1.11*a*). If four equivalent triangles meet at each apex the solid obtained is an octahedron with six vertices, eight faces, and twelve edges (Fig. 1.11*b*), while if five edges meet at a point an icosahedron is obtained (Fig. 1.11*c*) with twelve vertices, twenty faces, and thirty edges. If six equilateral triangles meet at a point then an infinite plane tessellation is obtained (Fig. 1.12*a*). Proceeding now with square faces: if three edges meet at a point we obtain the cube (Fig. 1.11*d*), and if four edges meet at a point we again obtain an infinite plane (Fig. 1.12*b*). Regular pentagons meeting three at a point yield the dodecahedron (Fig. 1.11*e*). As the angle of the regular pentagon is

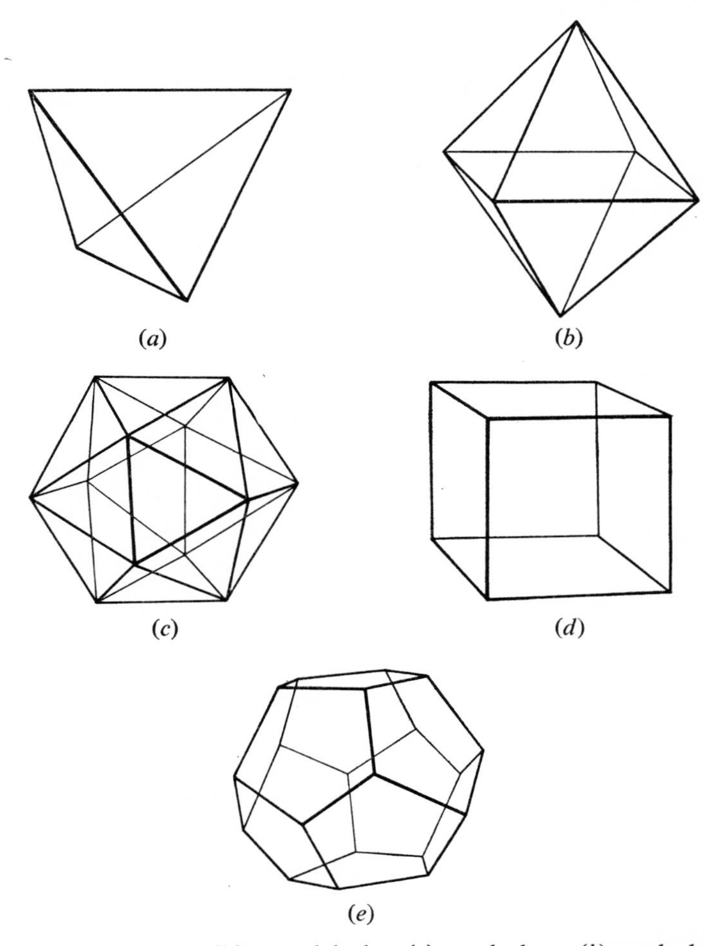

FIG. 1.11 The regular solids or polyhedra: (*a*) tetrahedron; (*b*) octahedron; (*c*) icosahedron; (*d*) cube; (*e*) dodecahedron.

greater than 90° it is clear that more than three pentagons cannot meet at a point. Hexagons can only form an infinite plane with three edges meeting at a point (hexagonal tessellation: Fig. 1.12*c*).

The properties of the five regular solids are given in Table 1.5 and are

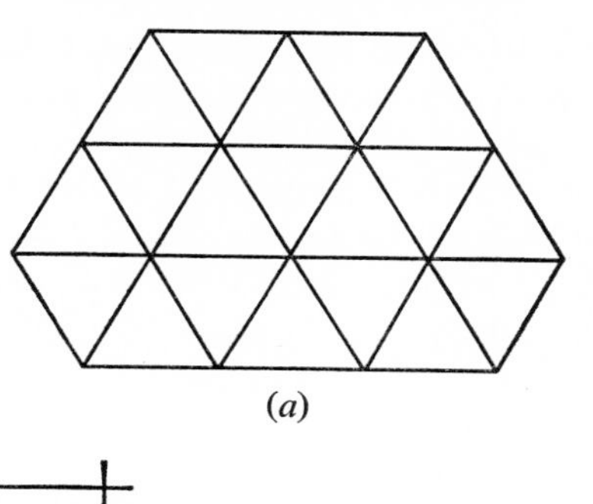

(*a*)

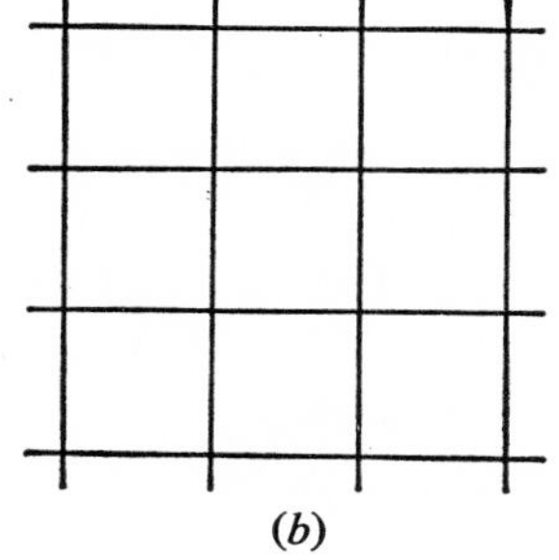

(*b*)

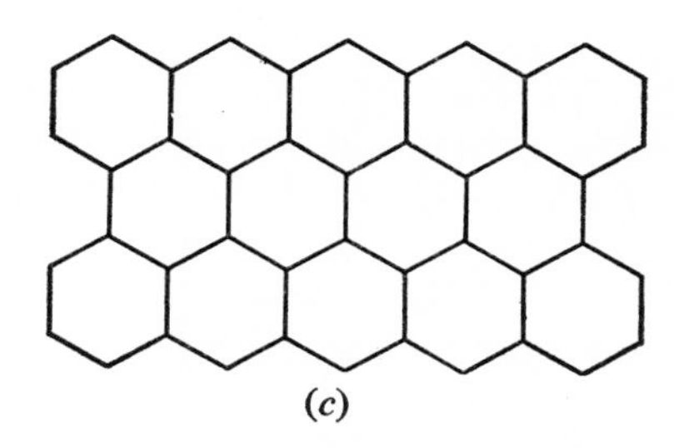

(*c*)

FIG. 1.12 Planar lattices or tessellation; (*a*) triangular tessellation; (*b*) square tessellation; (*c*) hexagonal tessellation.

summarized by the Descartes–Euler formula: number of vertices + number of faces — number of edges = 2.

The polyhedra that correspond to maximizing the least distance between points on a sphere all have a maximum number of triangular faces and therefore, of the regular solids, only the tetrahedron, the octahedron, and the icosahedron are found as arrangements of electron pairs, and the tetrahedral arrangement of four electron pairs and the octahedral arrangement of six electron pairs are by far the most common. Valence shells containing more than six

Table 1.5 The regular solids or polyhedra

Polyhedron	Faces	Number of edges	Number of faces	Number of vertices
Tetrahedron	triangles	6	4	4
Octahedron	triangles	12	8	6
Cube	squares	12	6	8
Dodecahedron	pentagons	30	12	20
Icosahedron	triangles	30	20	12

electron pairs are relatively uncommon and are discussed in Chapter 5. More than nine electron pairs in a valence shell is very uncommon, and therefore the icosahedral arrangement of twelve electron pairs, although known, is very rare. A number of other less regular polyhedra are discussed later in this book, and in every case the majority, if not all of their faces are triangular. We have already met the first of these polyhedra, namely, the trigonal bipyramid (Fig. 1.2), which has six triangular faces.

1.7 BOND PROPERTIES

A very large amount of information has been obtained in recent years by spectroscopic and diffraction techniques on the interatomic distances (bond lengths) in molecules and crystals. The length of a

Table 1.6 Covalent radii for some elements (Å)

Li	1·45	Na	1·80	K	2·20
Be	1·05	Mg	1·50	Ca	1·80
B	0·81	Al	1·25	Ga	1·30
C—	0·77	Si	1·17	Ge	1·22
C=	0·67	P—	1·10	As—	1·21
C≡	0·60	P=	1·00	As=	1·11
N—	0·70	S—	1·04	Se—	1·17
N=	0·62	S=	0·94	Se=	1·07
N≡	0·55	Cl	0·99	Br	1·14
O—	0·66	Ar	0·95	Kr	1·11
O=	0·62			Sb	1·41
O≡	0·55			Te	1·37
F	0·64			I	1·33
Ne	0·62				

The values are single bond radii except where otherwise indicated.
Sources: Linus Pauling, *Nature of the Chemical Bond*, 3rd Ed., Cornell University Press. J. C. Slater, *J. Chem. Phys.*, **41**, 3199 (1964). Values for the noble gases were obtained by extrapolation.

single bond between two given atoms very often varies rather little from one molecule to another and it is possible to divide up the bond distance between two such atoms into a contribution from each atom known as the *covalent radius* of the atom. To a good approximation these values are additive. Table 1.6 gives values for the covalent radii of a number of the elements taken from values given by Pauling and by Slater.

Various slightly different values of the covalent radii have been proposed by other authors and it should also be remembered that the covalent radius of an atom varies somewhat with its oxidation state and with the number and nature of the attached ligands. If these limitations are borne in mind the values are useful in that they can be used to give at least an approximate idea of the expected bond length for any single bond. Double bonds are in general shorter than single bonds and triple bonds are shorter still because of the greater

Table 1.7 Lengths of some multiple bonds (Å)

C—C 1·54	C—N 1·47	N—N 1·47	C—O 1·43	N—O 1·36
C=C 1·33	C=N 1·22	N=N 1·25	C=O 1·23	N=O 1·22
C≡C 1·21	C≡N 1·15	N≡N 1·09	C≡O 1·13	N≡O 1·06

attraction for the two positively charged cores exerted by the two electron pairs of a double bond or the three electron pairs of a triple bond. Some double bond and triple bond covalent radii are also given in Table 1.6 and some single, double, and triple bond lengths are compared in Table 1.7. A single bond is said to have a bond order of one, a double bond a bond order of two, and a triple bond a bond order of three. Bonds that have lengths that are intermediate between those of a single and a double bond or a double and a triple bond may be regarded as having a fractional bond order, an approximate value for which can be obtained from a bond-length–bond-order plot.

The enthalpy of dissociation (i.e. the heat of dissociation at constant pressure) of a diatomic molecule into its atoms is a useful measure of the strength of the bond, e.g.:

$$H_2 \rightarrow 2H \qquad \Delta H^\circ = 104 \text{ kcal mole}^{-1}$$
$$Cl_2 \rightarrow 2Cl \qquad \Delta H^\circ = 58 \text{ kcal mole}^{-1}.$$

This quantity is generally called the *bond energy*. For a polyatomic molecule containing more than one bond the heat of dissociation gives an average bond energy, e.g.:

$$H_2O \rightarrow 2H + O \qquad \Delta H = 220 \text{ kcal mole}^{-1}$$

giving an average bond energy for the OH bond of 110 kcal mole^{-1}. This value can also be used to a reasonable approximation for the OH bond in other molecules as it is found that the sum of such bond energies is usually quite close to the total dissociation energy of a molecule. Some average single bond energies are given in Table 1.8 and some bond energies for multiple bonds are given in Table 1.9.

Table 1.8 Some average single bond energies (kcal mole^{-1}) at 298°K (mostly those suggested by Pauling)

	Si	H	C	I	Br	Cl	N	O	F
F	129	135	105		61	61	65	44	38
O	88	110	84		57	49	53	33	
N		93	70			48	38		
Cl	86	103	79	50	52	58			
Br	69	87	66	43	46				
I	51	71	57	63					
C	69	99	83						
H	70	104							
Si	42								

Another useful measure of the strength of a bond is the *force constant* associated with its stretching vibration. For a diatomic molecule the frequency of vibration ν is related to the force constant k and the reduced mass $\mu = m_1m_2\ (m_1 + m_2)$, where m_1 and m_2 are the masses of the two atoms, by the equation

$$\nu = \frac{1}{2\pi}\sqrt{\frac{k}{\mu}}$$

Table 1.9 Bond energies for some multiple bonds (kcal mole^{-1})

C—C	N—N	C—O	C—N
82	38	84	70
C=C	N=N	C=O	C=N
147	100	170	147
C≡C	N≡N	C≡O	C≡N
194	226	262	210

The stronger the bond the higher the force constant and the higher the vibrational frequency. For a polyatomic molecule the stretching force constants for the various bonds cannot in general be obtained in a completely unambiguous and certain manner. Nevertheless the values obtained, although often approximate, are useful as an approximate measure of bond strength. Some typical values are given in Table 1.10.

Another useful bond property is the *dipole moment*. A diatomic molecule with a polar bond carries a positive charge at one end of the

Table 1.10 Force constants $10^5 k$ (dyne cm^{-1})

HF	8·83
HCl	4·81
HBr	3·84
HI	2·93
H—OH	7·7
H—SH	4·1
$H_3C—CH_3$	4·50
$H_2C=C_2H$	9·57
HC≡CH	15·72
HC≡N	17·7
C≡O	18·55
N≡N	22·40
OS=O	9·97
F_2	4·45
Cl_2	3·19
Br_2	2·42
I_2	1·70

Sources: G. Herzberg, *Infra-red and Roman Spectra of Polyatomic Molecules*, Van Nostrand, 1945. T. L. Cottrell, *The Strengths of Chemical Bonds*, 2nd Ed., Butterworths, 1958.

molecule and an equal negative charge at the other end. The product of the charge times the distance between the two charges is known as the dipole moment of the molecule. If the charge is measured in esu and the distance in cm then the dipole moment is given in Debye units (D) where 1 Debye = 10^{-18} esu. cm. The dipole moments of the hydrogen halides decrease in the series HF > HCl > HBr > HI indicating a decrease in bond polarity in this series consistent with the decreasing electronegativity of the halogen, Table 1.11.

Table 1.11 Dipole moments, bond distances and effective charges for the hydrogen halides

	Dipole moment μ(D)	Bond length (Å)	Effective charge* q_{eff}
HF	1·91	0·92	0·43
HCl	1·07	1·27	0·18
HBr	0·79	1·41	0·12
HI	0·38	1·61	0·05

* The effective charge is the charge in electron units (i.e., fractions of the charge on one electron) that is required when centred on the nuclei, separated by the observed internuclear distance, to give the observed dipole moment.

REFERENCES AND SUGGESTIONS FOR FURTHER READING

J. C. D. BRAND and J. SPEAKMAN, *Molecular Structure*, Arnold, London (1960).

F. A. COTTON and G. WILKINSON, *Advanced Inorganic Chemistry*, Interscience, 2nd Ed. (1966).

T. L. COTTRELL, *The Strength of Chemical Bonds*, Butterworths (1958).

H. S. M. COXETER, *Regular Polytopes*, 2nd Ed., Pitman Publishing Co. (1948).

R. J. GILLESPIE and R. S. NYHOLM, *Q. Rev. Chem. Soc.*, **11,** 339 (1957).

G. HERZBERG, *Infra-red and Raman Spectra of Polyatomic Molecules*, Van Nostrand (1945).

G. N. LEWIS, *J. Amer. Chem. Soc.*, **38,** 762 (1916).

L. PAULING, *Nature of the Chemical Bond*, 3rd Ed., Cornell University Press (1960).

N. V. SIDGWICK and H. E. POWELL, *Proc. Roy. Soc.*, **A 176,** 153 (1940).

J. C. SLATER, *J. Chem. Phys.*, **41,** 3199 (1964).

P. J. WHEATLEY, *The Determination of Molecular Structure*, Oxford, 2nd Ed. (1968).

2

The Pauli Exclusion Principle and Electron-Pair Orbitals

2.1 ELECTRON SPIN AND THE PAULI EXCLUSION PRINCIPLE

The justification for the rule proposed in Chapter 1 that electron pairs tend to adopt arrangements which maximize their distance apart is provided by the Pauli exclusion principle, which is a fundamental principle relating to the behaviour of electrons, which depends on the property known as spin. In addition to the properties of mass and charge, electrons have magnetic properties, i.e., they have a magnetic moment, and only two values of this magnetic moment are found which are of the same magnitude but of opposite sign. The magnetic moment of the electron can be thought of as arising from the spinning of the electron around its own axis, spin being possible in either a clockwise or an anti-clockwise sense, the moving electric charge of the electron thereby generating a magnetic moment in one direction or in the opposite direction. An electron is described by a wave-function ψ and the square of the wave-function ψ^2 measures the probability of finding the electron at some given point in space and with a given spin. This total wave-function is, to a good approximation, simply a product of a space wave-function and a spin wave-function.

It is a fundamental property of electrons that they conform to the exclusion principle. It is not possible to deduce the exclusion principle from any more fundamental principle, and it must be accepted as a fundamental fact concerning the behaviour of electrons. In its original form due to Pauli, it states that no two electrons can have the same space wave-function, i.e., occupy the same orbital, and have the same spin. It is another fundamental property of electrons that they are indistinguishable. Because of their indistinguishability the interchange of any two electrons in a system does not produce any observable change in the system. This means that the square of the wave-function must remain unchanged when the space and spin co-ordinates of two electrons are interchanged. This places a restriction on the wave-function itself; either it must remain unchanged, or it must change sign when the co-ordinates of any two electrons are interchanged, i.e., it must be either symmetrical or antisymmetrical to the interchange of the co-ordinates of any two electrons. Only antisymmetrical wave-functions are found to represent the properties of electrons, and this is in fact a more general statement of the exclusion principle, namely that the complete wave-function for any system must be antisymmetrical to electron interchange.

Thus, for the interchange of the co-ordinates (spatial and spin) of two electrons, 1 and 2 in any system, the Pauli principle states that

$$\psi(x_1, x_2, x_3 \ldots) = -\psi(x_2, x_1, x_3 \ldots),$$

where x_1 represents the co-ordinates for electron 1, etc. Now if two electrons have the same co-ordinates (spatial and spin), i.e., if $x_1 = x_2 = x$, then

$$\psi(x, x, x_3 \ldots) = -\psi(x, x, x_3 \ldots),$$

and hence

$$\psi(x, x, x_3 \ldots) = 0.$$

Thus, if two electrons have the same spin and space co-ordinates, the wave-function vanishes, or in other words, it is not possible for two electrons having the same spin to be at the same point in space. However, from the point of view of the Pauli exclusion principle there is no restriction on two electrons of opposite spin occupying the same point in space; although the space part of the wave-function is symmetrical to electron interchange in this case the spin part is antisymmetrical.

By extending the above argument somewhat, it can be shown that electrons having the same spin tend to keep apart and to occupy different regions of space, while electrons having opposite spin are allowed to come together, and indeed they tend to occupy the same region of space. The effect of the operation of the Pauli exclusion principle is quite separate and independent from the electrostatic repulsion between electrons that results from their negative charge. For electrons having the same spin, electrostatic repulsion reinforces their tendency to keep apart, whereas for electrons having opposite spin it opposes their tendency to come together. Thus the Pauli exclusion principle results in a correlation between the positions of the electrons in any system which may be called spin correlation and, in addition, there is electrostatic repulsion which gives rise to what may be called charge correlation. As a first approximation, it is reasonable to assume that for electrons of opposite spin the charge correlation approximately cancels the spin correlation, and there is little resulting correlation, but for electrons of the same spin, charge correlation reinforces spin correlation, and there is a strong tendency for electrons having the same spin to keep apart, i.e., to occupy different regions of space. Thus, if a pair of electrons of opposite spin are occupying a given region of space then there is only a very small probability that other electrons will be found in this region of space. In other words, a pair of electrons in a given region of space has a strong tendency to exclude other electrons from this space. A region of space which is primarily occupied by one pair of electrons may be called an orbital. Such a space has also been called a *loge* by Daudel.

Let us now consider the effect of the Pauli exclusion principle on the arrangement of the electrons in the outer or valence shell of an atom with completed spherical inner shells.

2.2 THE ARRANGEMENT OF ELECTRONS IN VALENCE SHELLS

The distribution of electrons in any system is determined by the kinetic energy of the electrons; by the electrostatic attractions exerted by the positively charged nuclei; by the electrostatic repulsions exerted by other electrons; and by the operation of the Pauli exclusion principle. The kinetic energy of an electron causes it to move around in, and to effectively occupy, as large a volume of space as the various restraints placed upon it by the nuclei and other electrons will allow. It is convenient to think of the electron as a charge cloud whose density at any point represents the probability of finding the electron at that point. For a system consisting of one nucleus and a single electron, there is no reason to expect that the electron will be more likely to be found in any one direction from the nucleus than in any other. Such an electron distribution is spherical, i.e., it has no angular dependence. Thus, for the simple system of a nucleus and a single electron, the electron may be represented by a spherical charge cloud. The probability of finding the electron at great distances from the nucleus is very small, and therefore the electron density becomes negligible, and an arbitrary spherical surface may be drawn which will then effectively contain all the electron. This spherical space occupied by the electron may be described as an orbital, and in this case it is a spherical or *s* orbital. When a second electron is added to this system it is attracted by the nuclear charge, and if it is of opposite spin to the first electron it tends to occupy the same spherical region of space around the nucleus, because, in accordance with the Pauli exclusion principle, two electrons of opposite spin are allowed to come close together, and although there is an electrostatic repulsion between them the two electrons may be regarded, to a reasonable approximation, as occupying the same spherical orbital. These two electrons constitute the K shell which thus contains only one orbital. On the addition of a third electron this must occupy a region of space outside the K shell, i.e., it occupies an orbital in the second, or L, shell and in the absence of other electrons it is reasonable to suppose that this orbital will also be spherical and will surround the nucleus and the inner 1*s* electron pair—this is the 2*s* orbital. A fourth electron enters the 2*s*

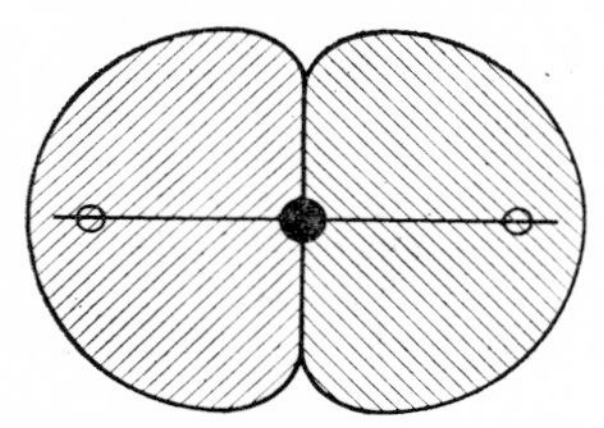

FIG. 2.1 Linear arrangement and approximate orbitals for two electrons with the same spin in the same valence shell.

orbital if it has opposite spin, but if it has the same spin as the third electron it cannot occupy the same space, and assuming that it remains in the same shell and therefore at approximately the same distance from the nucleus the two electrons keep apart having a most probable location with respect to each other of 180°. The two electrons may be imagined as sharing out the space around the central core into two approximately hemispherical regions or orbitals (Fig. 2.1). This does not imply that in the free atom the total electron distribution is non-spherical, because in the absence of some fixed direction in space any orientation of the two electrons with respect to a fixed axis is possible, only their relative orientation tends to be at 180°.

For three electrons with the same spin, the distribution which keeps them as far apart as possible is clearly that in which the three electrons lie at the corners of an equilateral triangle, surrounding the central core and making angles of 120° with each other at the central core (Fig. 2.2). For four electrons with the same spin, the arrange-

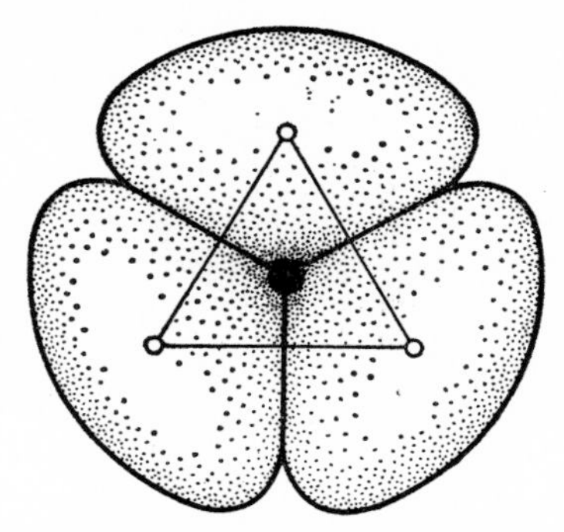

FIG. 2.2 Planar triangular arrangement for three electrons with the same spin in the same valence shell.

ment which keeps them as far apart as possible is a tetrahedral arrangement with the central core at the centre of the tetrahedron (Fig. 2.3). Figure 2.4 gives another representation of how two, three, and four electrons of parallel spin share out the space around the

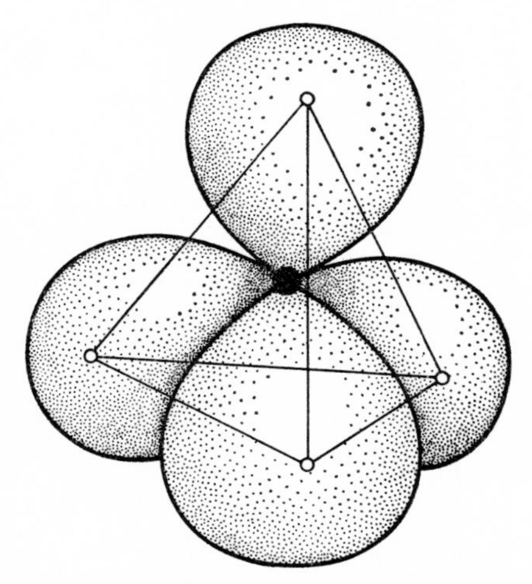

FIG. 2.3 Tetrahedral arrangement and approximate orbitals for four electrons with the same spin in the same valence shell.

central core of an atom, each occupying a segment of a sphere which represents the orbital for this electron. There is no need to continue this somewhat hypothetical discussion of free atoms with valence shells containing electrons of the same spin, but we will now consider a situation of much more relevance to the discussion of molecular shape, namely, that in which there are equal numbers of electrons of

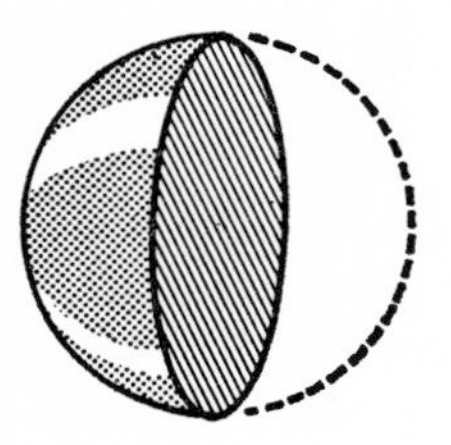
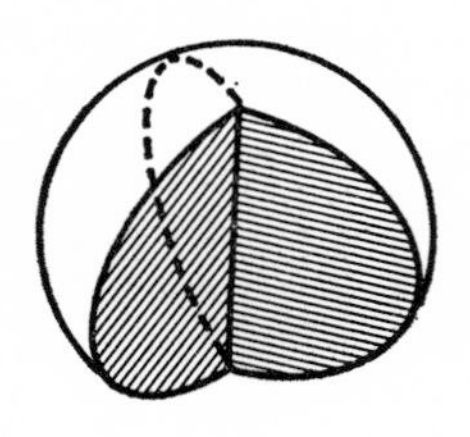
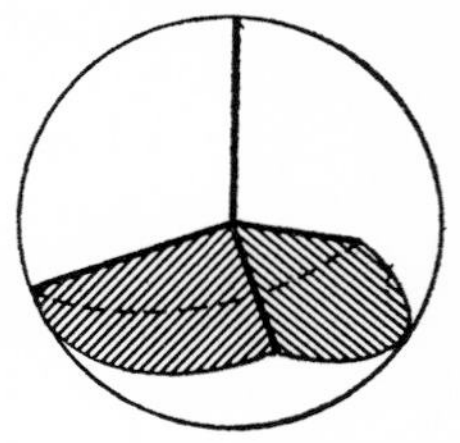

FIG. 2.4 Segments of a sphere representing orbitals for two, three, and four electrons with the same spin.

opposite spin—which is the case in the vast majority of stable molecules.

Consider a valence shell containing six electrons: three of one spin and three of opposite spin. As a consequence of spin correlation the most probable arrangement of the electrons in each spin set is triangular (Fig. 2.5). In each individual set the tendency towards a

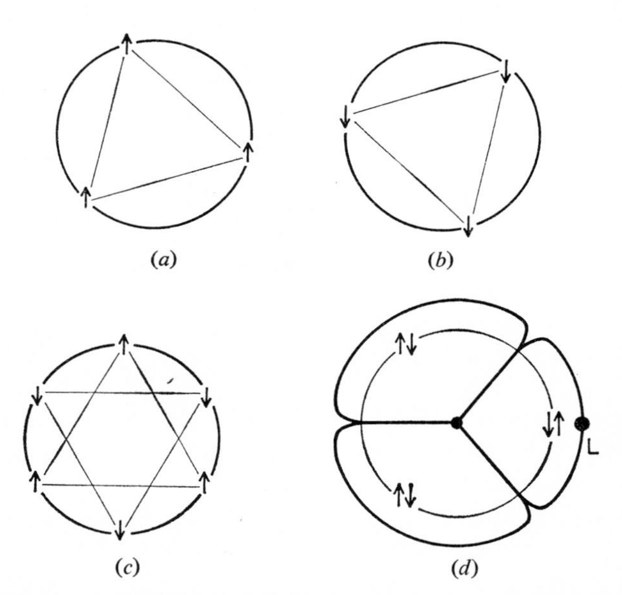

FIG. 2.5 (*a*) Set of three electrons with the same spin confined to a circle. (*b*) Another set of three electrons with opposite spin to those in (*a*). (*c*) Both sets of three electrons confined to the same circle. Spin correlation maintains the most probable triangular arrangement in each set but the two sets tend to keep apart as the result of electrostatic repulsion. (*d*) In the presence of a ligand 'L' which causes two electrons of opposite spin to occupy the same bonding orbital the electrons become grouped in pairs of opposite spin.

triangular arrangement is increased by the electrostatic repulsion or charge correlation between the electrons. For electrons of opposite spin, spin correlation tends to bring the electrons together; but charge correlation tends to keep them apart, so that there is little resultant correlation between electrons of opposite spin, and thus the two triangular groups of electrons can be regarded to a first approximation as being independent of each other. However, in a molecule, the electrons in the valence shell of an atom are always under the influence of at least one other nucleus, and this necessarily introduces a correlation between the two sets of electrons of opposite spin. In the case under consideration, two electrons of opposite spin are drawn into the same region of space to form a bond with a second nucleus and this automatically draws the remaining electrons into close coincidence. Thus, in the valence shell of an atom containing three electrons of one spin and three electrons of opposite spin, and which is forming at least one bond, there will be three pairs of electrons arranged in a plane at 120° to each other.

For the important case of an atom with eight electrons (four of one spin and four of opposite spin) in its valence shell there will be two sets of four electrons of opposite spin: each with a tetrahedral arrangement. The formation of a bond will cause these two tetrahedra to come into partial coincidence, although rotation of the two sets with respect to each other is still possible around the bond axis. The formation of a second bond then brings the two tetrahedra into close coincidence, four close pairs of electrons surrounding the central core in a tetrahedral arrangement (Fig. 2.6).

We see then, that the Pauli exclusion principle provides the justification for the assumption that the valence shell of an atom in a molecule contains pairs of electrons that have a most probable arrangement that keeps them as far apart as possible. However, it may be noted that for diatomic molecules having cylindrical symmetry around the bond there is no requirement that the electrons located around the bond axis should form close pairs.

Because of its kinetic energy each electron spreads out as much as it is allowed to by the nuclei and other electrons. The electron pairs can thus be regarded as sharing out the available space around the central core so that each pair of electrons occupies a region of space or orbital centred upon the vertices of the polyhedra in Table 1.3. The Pauli exclusion principle allows each orbital to be occupied by two electrons of opposite spin but, as it keeps electrons of the same spin apart, it does not allow other electrons to enter a region of space already occupied by two electrons. We may think of each pair

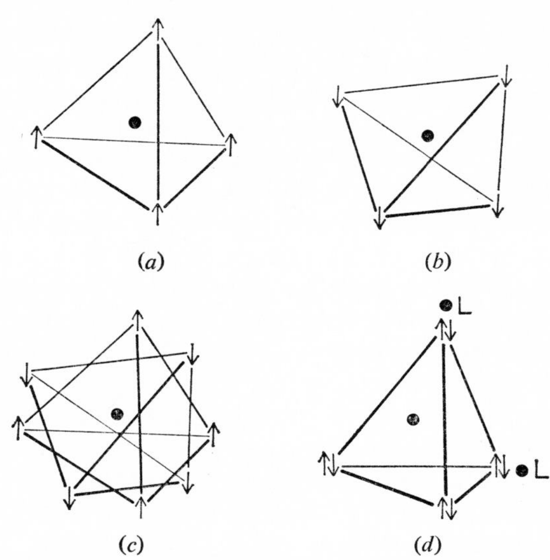

FIG. 2.6 (*a*) and (*b*) Sets of four electrons with the same spin. (*c*) Two sets of four electrons of opposite spin occupying the same valence shell. (*d*) Four close pairs of electrons resulting from the presence of the two ligands 'L'.

of electrons as a more or less impenetrable charge cloud which takes up a certain amount of space and excludes other electrons from this space, i.e., each charge cloud may be regarded as repelling other charge clouds.

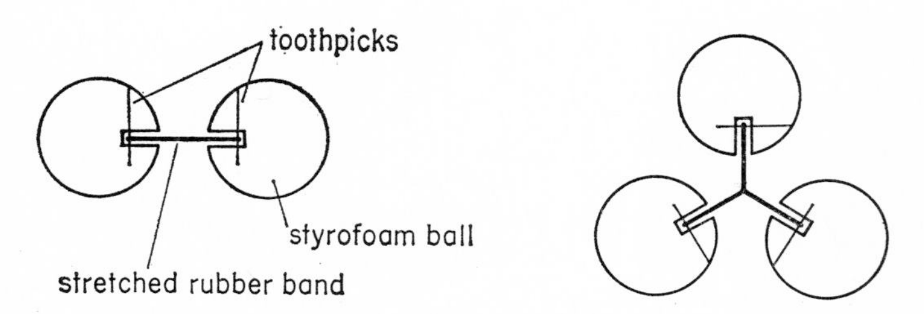

FIG. 2.7 Illustration of the construction of models for illustrating electron-pair arrangements according to the hard-sphere model.

The possible arrangements of different numbers of electron pairs can then be demonstrated by means of a rather simple model. The charge cloud corresponding to each electron pair is represented by a spherical polystyrene (styrofoam) ball. Pairs of balls are held together by a stretched rubber band that is held inside each ball by means of a toothpick (Fig. 2.7). Clusters of four and six balls are then formed by twisting the rubber bands of two or three pairs together. These clusters will be found to adopt the tetrahedral and octahedral arrangements respectively (Fig. 2.8). The rubber bands represent the attraction of the central positive core of the atom for the electron clouds. A group of three balls can also be constructed as shown in Fig. 2.7, and when a cluster of five is formed with a pair, the trigonal bipyramid arrangement results (Fig. 2.8). These models can be used

FIG. 2.8 The hard-sphere models for the tetrahedral, trigonal bipyramidal and octahedral arrangements of four, five and six electron pairs respectively.

to demonstrate that other arrangements, e.g., the square arrangement for four balls or the square pyramid for five, are less stable. By forcing the balls into these arrangements and then gently shaking the model rearrangement to the most stable form occurs.

The assumption that the charge cloud of an electron pair can be represented by an impenetrable sphere provides a very useful and apparently reasonably accurate model which is used considerably in this book. This model was first proposed by Kimball and has been considerably developed by Bent who has called it the tangent-sphere model. It may also be called the hard-sphere model.

2.3 OVERLAP AND INTERPENETRATION OF CHARGE CLOUDS

In the simple model discussed above, electron charge clouds have been considered as hard, impenetrable spheres which have no interaction with each other if they do not touch, but repel each other with an infinite force as soon as they touch. This corresponds to a force law of the type $F = 1/r^n$ where r is the distance between the surfaces of the two charge clouds and n approaches infinity. It would be more realistic to consider that charge clouds do not have a definite well-defined surface, and that they can penetrate each other and deform each other to some extent, although such overlap and interpenetration is strongly resisted by the operation of the Pauli exclusion principle. This can be represented to a reasonable approximation by assuming that charge clouds repel each other in accordance with a force law of the type $F = 1/r^n$, where n is a large number, probably of the order of 10, and where r is the distance between the 'centres' of the charge clouds, i.e., the point of maximum electron density. If the force between charge clouds was purely electrostatic then the appropriate force law would be Coulomb's law, i.e., $F = 1/r^2$. If we use the model of points on the surface of a sphere, each point representing an electron pair, and assume Coulomb's law of repulsion between the points, then it has been shown that for two to six points the arrangements of points which minimize the energy are the same as the arrangements which maximize the least distance between any two particles. Maximizing the least distance between any two particles is equivalent to a force law of the type $F = 1/r^n$ where n approaches infinity. Thus, although we have no knowledge of the best value to use for n, it would seem to be a reasonable assumption that the arrangement of any given number of particles up to six that minimizes the energy is independent of n since the same arrangement is obtained for $n = 2$ and for $n = \infty$. Hence we conclude that the most probable arrangements of electron pairs given above are valid for soft charge clouds which can deform each other and penetrate into each other to some extent, as well as for hard impenetrable charge clouds. The interaction or repulsion between these charge clouds then arises when they overlap each other, as this overlap is resisted by the operation of the exclusion principle, and with increasing overlap the repulsive force between two orbitals increases rapidly.

2.4 THE SIZES OF ELECTRON-PAIR ORBITALS

Since the structures of molecules depend on the arrangements of electron pairs in valence shells it is important to have some information on the sizes of electron pairs and on the number that can occupy a given valence shell. A rough approximation to the size of an electron pair can be obtained using the hard-sphere model. The length of a covalent bond, d, can then be taken to be equal to the sum of the inner core radii of the two atoms plus the diameter of the shared electron pair.

$$d = r^{A}_{core} + r^{B}_{core} + 2r_e,$$

where r_e is the radius of the shared electron pair (Fig. 2.9). Since,

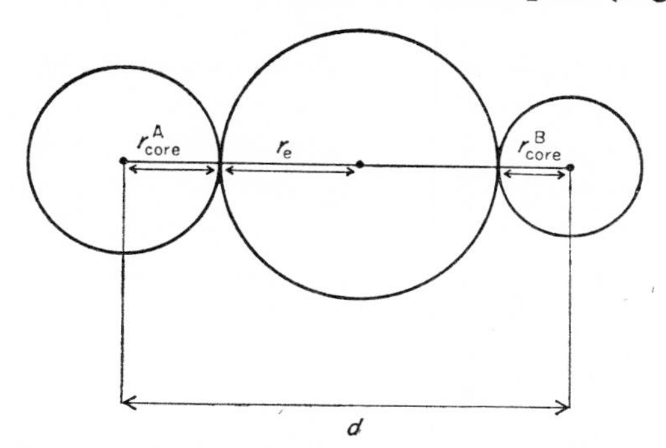

FIG. 2.9 Relation of the radius of an electron-pair orbital r_e to the covalent bond length d.

for a homonuclear diatomic molecule the covalent radius $r_{covalent} = \frac{1}{2}d$ we have

$$r_{covalent} = r_{core} + r_e$$

or

$$r_e = r_{covalent} - r_{core}.$$

Using covalent radii from Table 1.6 and ionic core radii given by Pauling, and listed in Table 2.1, values of the electron pair radii may be calculated for most of the main group elements: these are also given in Table 2.1.

For the elements in each row of the periodic table the radius of a bonding electron pair tends to reach approximate limiting values of 0·6 Å for the elements up to neon, 0·75 Å for the elements up to argon, 0·75 Å for the elements up to krypton and 0·8 Å for the elements up to xenon. Thus all shared electron pairs are seen to be of approximately the same size but are somewhat smaller when attracted by the very small, highly charged inner cores of the elements boron to neon. Knowing the size of the inner core of an atom and the size of the electron pairs in its valence shell, it is possible to calculate how many pairs can be packed around the inner core and thus to determine the maximum size of the valence shell, i.e., the co-ordination number of the inner core for electron pairs. This can be done very simply by using the radius ratio criterion that is used to determine the co-ordination number for the packing of anions around a cation in a crystal. The limiting ratios of r_{core}/r_e for each possible co-ordination number are given in Table 2.2; a given co-ordination number is only possible (unless the electron pairs are compressed or are not in contact with the nucleus) when a given radius ratio is exceeded. In this way the co-ordination numbers given in Table 2.1

Table 2.1 Bonding electron-pair radii (A)

	$r_{covalent}$	r_{core}	r_e	r/r_e	Co-ordination number
Li	1·45	0·60	0·85	0·71	8
Be	1·05	0·31	0·74	0·42	4–6
B	0·81	0·20	0·61	0·33	4
C	0·77	0·15	0·62	0·24	4
N	0·70	0·11	0·59	0·19	3
O	0·66	0·09	0·57	0·16	3
F	0·64	0·07	0·57	0·12	2
Ne	0·62	0·06	0·56	0·11	2
Na	1·80	0·95	0·85	1·12	>9
Mg	1·50	0·65	0·85	0·77	9
Al	1·25	0·50	0·75	0·67	8
Si	1·17	0·41	0·76	0·54	6
P	1·10	0·34	0·76	0·45	6
S	1·04	0·29	0·75	0·39	4–6
Cl	0·99	0·26	0·73	0·36	4
Ar	0·95	0·23	0·72	0·32	4
K	2·20	1·33	0·87	1·53	>9
Ca	1·80	0·49	0·81	1·22	>9
Ga	1·30	0·62	0·68	0·91	9
Ge	1·22	0·53	0·69	0·77	9
As	1·21	0·47	0·74	0·64	7–8
Se	1·17	0·42	0·75	0·56	6
Br	1·14	0·39	0·75	0·52	6
Kr	1·11	0·36	0·75	0·48	6
Sb	1·41	0·62	0·79	0·78	9
Te	1·37	0·56	0·81	0·69	8
I	1·33	0·50	0·83	0·60	6–7

were obtained. The maximum co-ordination number decreases from eight for lithium, to four for boron and carbon, to three for nitrogen and oxygen, and to two for fluorine. These results appear at first sight to be somewhat unsatisfactory for nitrogen, oxygen, and fluorine, which are known to have four electron pairs in their valence shells almost without exception. We can only conclude that these electron pairs are compressed rather strongly and in addition, or alternatively, they are forced away from the central core by mutual repulsion, and in the hard-sphere model would no longer be touching the core. The situation can be compared to that of a small cation, e.g., Li^+ which does not fill the tetrahedral or octahedral hole in a close-packed array of anions. As we shall see, this leads to a number of very important consequences, and is one reason why the chemistry of the elements carbon to fluorine differs so much from the heavier elements in the respective groups of the periodic table.

The maximum co-ordination numbers calculated for the second period (Na to Ar) clearly show that the valence shells of these elements

Table 2.2 Minimum radius ratios

Polyhedron	Co-ordination number	Minimum radius ratio
Triangle	3	
Tetrahedron	4	0·225
Octahedron	6	0·414
Monocapped octahedron	7	0·592
Square antiprism	8	0·645
Tricapped trigonal prism	9	0·732

are larger than for those of the preceding elements. Following the large values for the metallic elements we find values of six for silicon, phosphorus, and sulphur. The size of the sulphur core is just less than that needed for six co-ordination but, bearing in mind the possibility of slight compression of the electron pairs and the approximate nature of the calculated sizes of the electron pairs, it is reasonable to conclude that sulphur might be able to achieve a co-ordination number of six. The predicted values for chlorine and argon are only four. Higher numbers of electron pairs are sometimes found in the valence shell of chlorine when it is attached to very electronegative ligands: this is discussed in the next chapter. For the next period, the co-ordination number for electron pairs decreases from large values for the metallic elements to six for selenium, bromine, and krypton. In the following period the values are still larger, decreasing to eight for tellurium and seven for iodine and six for xenon. These co-ordination numbers can only be regarded as approximate, and can be modified somewhat when account is taken of the varying sizes of electron pairs as discussed in the following chapter. The values do however agree reasonably well with the observed co-ordination numbers of the main group non-metallic elements and with the general increase in maximum co-ordination number that is observed with increasing atomic number in any group of the periodic table, and of course with the decrease that is observed with increasing atomic number in any period of the periodic table.

REFERENCES AND SUGGESTIONS FOR FURTHER READING

H. A. BENT. *J. Chem. Ed.*, **40,** 446, 523 (1963); **42,** 302, 348 (1965); **44,** 512 (1967); **45,** 768 (1968).

R. DAUDEL. *The Fundamentals of Theoretical Chemistry.* Pergamon, 1968.

G. E. KIMBALL. References to unpublished work by Kimball and his students are given by Bent.

J. W. LINNETT. *Wave Mechanics and Valency.* Methuen, 1960.
The Electronic Structure of Molecules. Methuen, 1964.

J. E. LENNARD-JONES. *Adv. Sci.*, **51,** 136 (1954).

L. PAULING. *Nature of the Chemical Bond*, 3rd Ed., Cornell University Press, 1960.

3

The Effects of Non-Equivalence of Electron Pairs

In Chapter 1 it was assumed that the electron pairs in a given valence shell are all equivalent irrespective of whether they are bonding or non-bonding pairs and, if they are bonding pairs, of the nature of the ligand that they are bonding. In fact this is not the case, and the shapes of molecules given in Chapter 1 are only correct to a first approximation. Generally there are small deviations from these predicted shapes, e.g., the bond angles in NH_3 and H_2O are 107·3° and 104·5° respectively, and are thus smaller than the predicted tetrahedral angle of 109·5°; the bond angle in ClF_3 is 87·5°, which is again smaller than the predicted angle of 90°. These deviations from the ideal angles are a consequence of the fact that non-bonding electron pairs are not equivalent to bonding pairs. The molecular shapes predicted in Chapter 1 are only exactly correct when all the electron pairs are binding identical ligands as, for example, in CH_4 and SF_6. In all other cases, e.g., when there are one or more non-bonding pairs of electrons, or when there are two or more different ligands, deviations from these ideal shapes are observed. These arise because the electron pairs in a valence shell do not, in general, all have the same size and shape and do not have equal interactions with each other.

3.1 NON-BONDING OR LONE-PAIRS OF ELECTRONS AND BOND ANGLES

Since a non-bonding pair of electrons is under the influence of only one positive core it is expected to be somewhat larger than a bonding pair in the same valence shell as the bonding pair is in the field of two positive cores. An approximate estimate of the size of non-bonding pairs can be obtained from the radii of negative ions such as P^{3-}, S^{2-}, and Cl^- which all have valence shells of four non-bonding pairs of electrons. Using the simple hard-sphere model the radius of a non-bonding electron pair is given by $r_e = \frac{1}{2}(r_{ion} - r_{core})$ as shown in Fig. 3.1. Some values obtained in this way are given in Table 3.1,

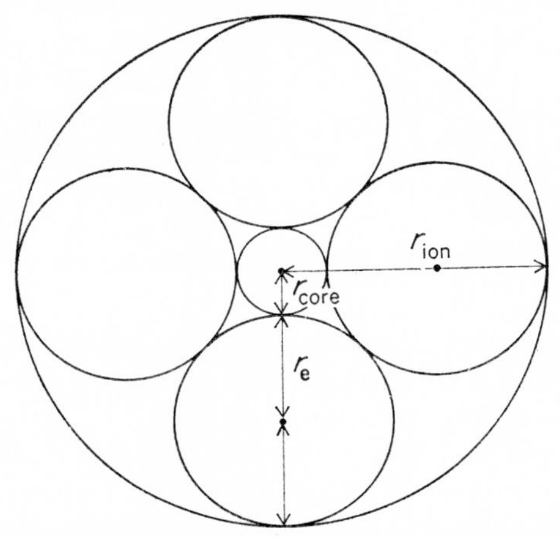

FIG. 3.1 Estimation of the size of non-bonding (lone) pairs from the ionic radius on the basis of the hard sphere model.

Table 3.1 Non-bonding electron-pair radii (Å)

	r_{ion}	r_{core}	r_e	r_{cor}/r_e	Co-ordination number
C	2·60	0·15	1·17	0·13	3
N	1·71	0·11	0·80	0·14	3
O	1·40	0·09	0·66	0·14	3
F	1·36	0·07	0·65	0·11	2
Ne	1·32	0·06	0·63	0·10	2
Si	2·71	0·41	1·15	0·36	4
P	2·12	0·34	0·89	0·38	4
S	1·84	0·29	0·77	0·38	4
Cl	1·81	0·26	0·77	0·34	4
Ar	1·78	0·23	0·77	0·30	4
Ge	2·72	0·53	1·15	0·46	6
As	2·22	0·47	0·87	0·54	6
Se	1·98	0·42	0·78	0·54	6
Br	1·93	0·39	0·78	0·50	6
Kr	1·88	0·36	0·76	0·47	6

and it may be seen that these non-bonding electron-pair radii are indeed somewhat larger than the bonding electron-pair radii given in Table 2.1 and consequently they lead to smaller co-ordination numbers than were given there. These are consistently less than four for the elements carbon to neon, equal to four for the elements silicon to argon and equal to six for the elements germanium to krypton.

Since a non-bonding electron pair is under the influence of only one atomic core it will tend to surround this core to the maximum possible extent. In the absence of any other electron pairs around the core it would occupy a spherical orbital symmetrically located around the inner core. However, in the presence of other electron

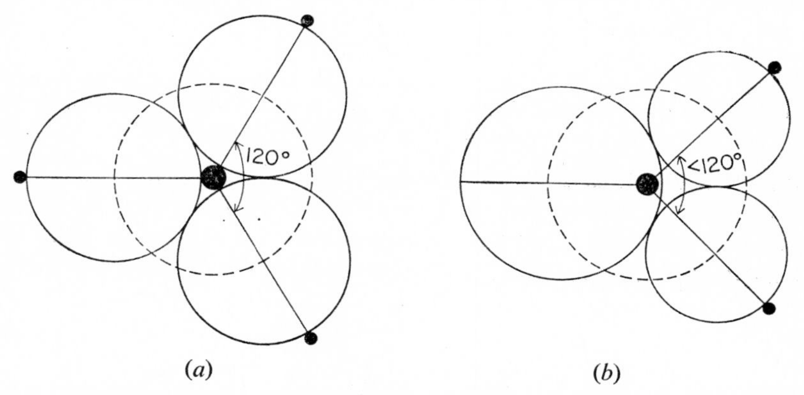

FIG. 3.2 Illustration of the effect of large lone-pairs on bond angles: (*a*) symmetrical planar arrangement of three equivalent bonding electron pairs with a bond angle of 120°; (*b*) Unsymmetrical arrangement of two bonding pairs and one lone-pair leading to a bond angle of less than 120°.

pairs, the extent to which it can surround the inner core will be limited. This is illustrated in Fig. 3.2 for a planar arrangement of three pairs of electrons. It may be seen that the larger size of the non-bonding electron pair and its more symmetrical position with respect to the central inner core both lead to a decrease in the angle between the bonding pairs. The more symmetrical position of the non-bonding pair with respect to the core also means that it occupies more of the surface of the central atom than a non-bonding pair. This effect is accentuated if the lone-pair is also somewhat larger than a bonding pair. Finally, if we modify the hard-sphere model to allow some distortion of the electron pairs—in particular to allow the non-bonding pairs to spread out to cover as much of the surface of the atom as is available to them—we arrive at a picture for the non-bonding and bonding electron pairs in a valence shell such as that given in Fig. 3.3.

Because lone-pairs of electrons tend to occupy more of the surface of an atom, and are in general larger, more spread out, and more

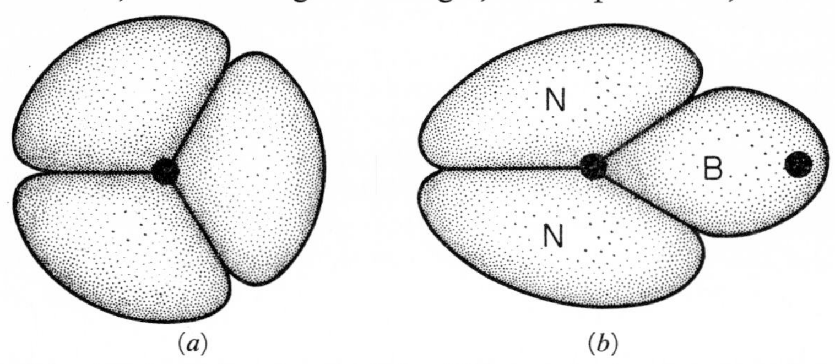

FIG. 3.3 An alternative picture of the differences in size and shape between bonding and non-bonding electron pairs: (*a*) three equivalent bonding pairs; (*b*) two non-bonding pairs N and a bonding pair B.

symmetrically located with respect to the inner core, the angles between bonding pairs in the same valence shell are less than they would be between equivalent electron pairs as shown in Fig. 3.2. Consequently bond angles are smaller than the ideal values associated

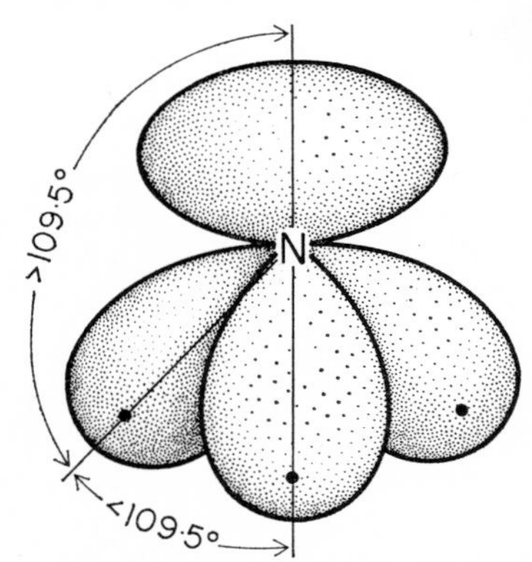

FIG. 3.4 Bonding and non-bonding electron pair orbitals in the ammonia molecule. The larger size of the non-bonding electron pair causes the angle between the bonding pairs to be less than the ideal angle of 109·5° for a regular tetrahedral molecule.

with a given number of equivalent electron pairs, i.e., 109·5° for a tetrahedral arrangement of four pairs and 90° for an octahedral arrangement of six pairs. Thus in the series CH_4, NH_3, and H_2O the bond angle decreases from 109·5° to 107·3° and to 104·5° as the number of non-bonding pairs increases (Fig. 3.4). Since an unshared pair of electrons tends to spread out and occupy more space around the central core than the bonding pairs, it is reasonable to assume that it interacts with other electron pairs, i.e., repels them, more strongly than the bonding pairs would. Thus we arrive at the general conclusion that unshared or lone-pairs of electrons are larger than bonding pairs in the same valence shell and that they interact with and repel other electron pairs more strongly than bonding pairs. The greatest interaction would be expected to occur between two lone-pairs and therefore we can make the useful generalization that electron-pair repulsions decrease in magnitude in the following order:

Lone-pair : lone-pair (l : l) > bond-pair : lone-pair (b : l) > bond-pair : bond-pair (b : b)

3.2 LONE-PAIRS IN OCTAHEDRAL AND TRIGONAL BIPYRAMID VALENCE SHELLS

If alternative positions are available in a valence shell for one or more lone-pairs then the lone-pairs have a strong tendency to occupy those positions in which the largest amount of space is available and in which their interactions with other electron pairs are therefore minimized. Since the interaction between lone-pairs is the strongest

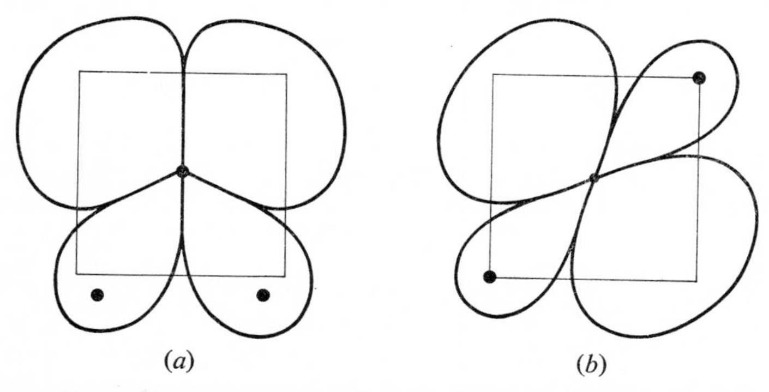

FIG. 3.5 Alternative arrangements for two bonding electron pairs and two non-bonding electron pairs in one plane of an octahedral AX_4E_2 molecule: (*a*) *cis* lone-pairs, (*b*) *trans* lone-pairs.

of the interactions between electron pairs, then it might be reasonably concluded that lone-pairs would tend to keep as far apart as possible. Hence in an AX_4E_2 molecule, e.g., XeF_4, they occupy *trans* positions (Fig. 1.8).

This is an important general conclusion that merits further discussion. If we consider the plane containing the two lone-pairs in an

AX_4E_2 molecule, the two possible arrangements of the electron pairs are as shown in Fig. 3.5. We see that the *trans* arrangement involves four b–l repulsions whereas the *cis* arrangement involves one b–b, two b–l and one l–l repulsion. Thus the condition that the *trans* arrangement is more stable is that

one b–b repulsion + one l–l repulsion > two b–l repulsions.

Since the repulsive force between two electron pairs varies inversely as some high power of the distance between them, the interaction energy as a function of the separation of any two electron pairs, r, may then be represented approximately as in Fig. 3.6, in which we see that because of their larger size the energy of interaction between lone-pairs begins to increase rapidly at larger distances r than

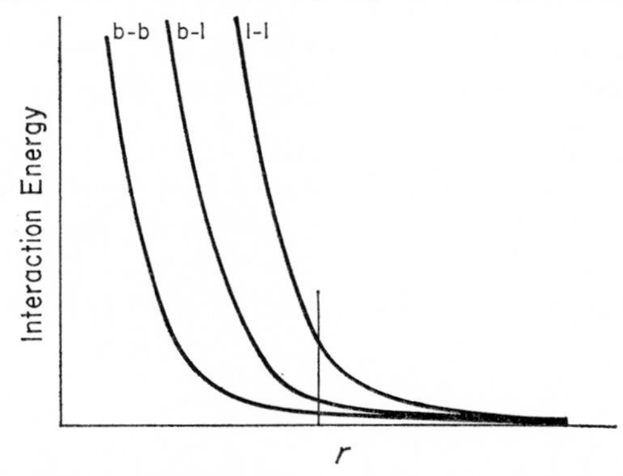

FIG. 3.6 Approximate representation of the dependence of the interaction energy between electron pairs on the distance between the electron pairs r. b–b, bond-pair : bond-pair interaction; b–l, bond-pair : lone-pair interaction; l–l, lone-pair : lone-pair interaction.

for b–l and l–l interactions. So at many distances r, such as that shown in Fig. 3.6, the above condition is satisfied.

The trigonal bipyramid arrangement of five electron pairs differs from the predicted arrangements of three, four, and six electron pairs in that the five pairs are not equivalent. Indeed, with the exception of the pentagonal plane, which clearly does not maximize distances between the points, there is no way in which five points can be arranged on the surface of a sphere so that they are all equivalent. In a trigonal bipyramid the two axial positions are not geometrically equivalent to the three equatorial positions. The axial positions have three nearest neighbours at 90°, namely the three equatorial positions, while an equatorial position has only two nearest neighbours at 90° —the two axial positions—and two other neighbours at a greater distance—the two equatorial positions at 120°. This has a nu important and interesting consequences which are discusseo (Chapter 4). At the moment we note only that we would expect lone-pairs would occupy those positions in which there is most roo for them and in which their interactions with other electron pairs are minimized. This is clearly the equatorial positions in the trigonal bipyramid. An electron pair situated at one of these positions does not have equivalent interactions with its neighbours, and assuming that the repulsion between two electron pairs decreases very rapidly with their distance apart, then it is only necessary to consider interactions between electron pairs that are at the smallest distance apart. Thus lone-pairs occupy equatorial positions since they then have only

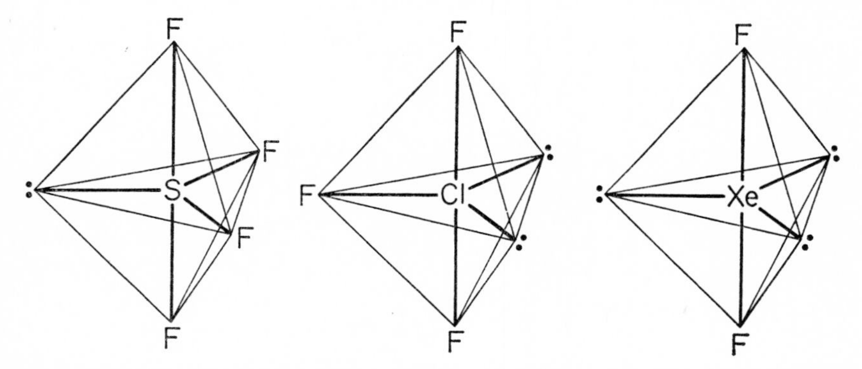

FIG. 3.7 Structures of SF_4, ClF_3 and XeF_2 illustrating the fact that lone-pairs always occupy the equatorial positions in a trigonal bipyramidal arrangement of five electron pairs.

two neighbouring electron pairs at 90°, while in the axial position they would have three neighbours at 90°. In all known cases, lone-pairs occupy the equatorial positions in a trigonal bipyramid arrangement, giving the structures of SF_4, ClF_3, and XeF_2 for one, two, and three lone-pairs respectively (Fig. 3.7).

The presence of a large lone-pair in an octahedral or trigonal bipyramid valence shell produces similar deviations from the ideal angles of a regular structure to those we have discussed previously for tetrahedral valence shells. For example, in BrF_5, SF_4, and ClF_3 the bond angles are less than ideal values of 90° and 180° (Fig. 3.8).

There is another interesting consequence of introducing a lone-pair into an octahedral arrangement which arises because, unlike the trigonal planar and tetrahedral arrangements, the lone-pair does not interact equally with all the remaining bond-pairs, and consequently

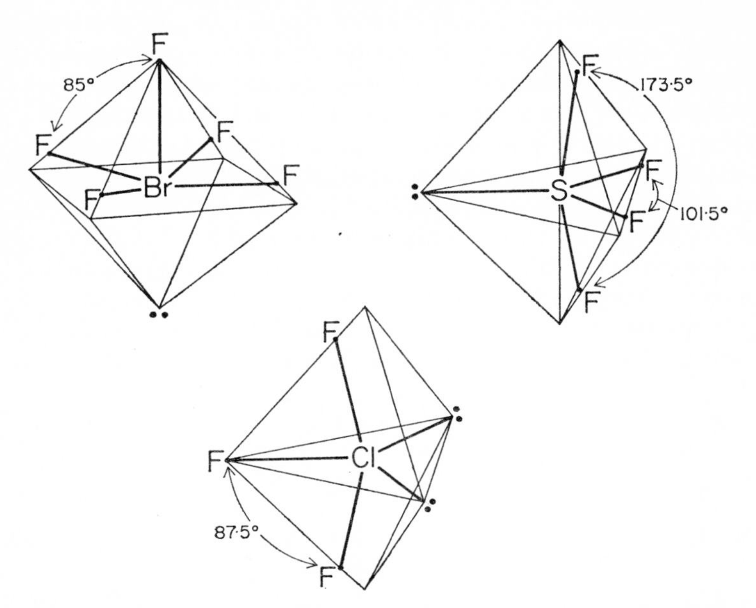

FIG. 3.8 Structures of BrF_5, SF_4 and ClF_3 illustrating the small deviations from the ideal shapes of AX_5E, AX_4E, and AX_3F_2 molecules produced by the lone-pairs.

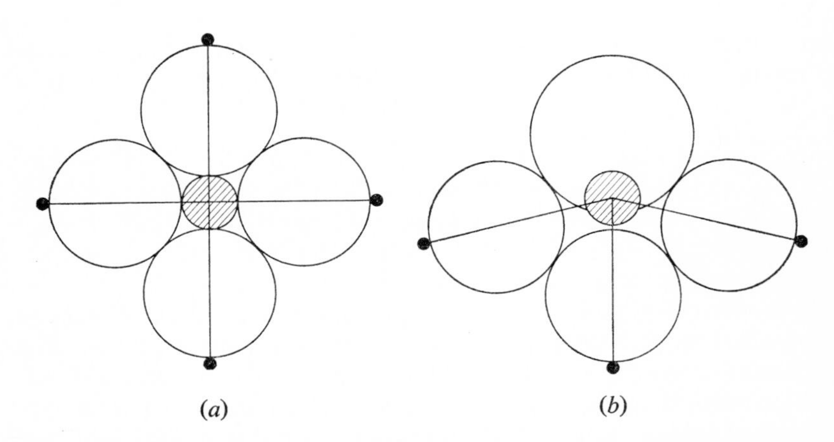

FIG. 3.9 Section through an octahedral arrangement of six electron pairs illustrating the effect of a non-bonding pair on bond lengths and bond angles: (*a*) six equivalent bonding pairs; (*b*) five bonding pairs and one non-bonding pair. The four bonding pairs adjacent to the lone-pair are pushed further away from the central core than the single *trans* bonding pair.

does not affect them all equally. In such an octahedral arrangement, in order to accommodate the larger lone-pair in an octahedral arrangement of six electron pairs, not only the bond angles but also the bond lengths are distorted. The distortion of bond lengths is not apparent in the trigonal planar and tetrahedral arrangements because all the bonds are affected equally, but this is not the case for the octahedral arrangement. Figure 3.9 shows how the replacement of an electron pair by a lone pair in an octahedral arrangement of six equivalent pairs produces not only a decrease in the angle between the bonding pairs but an increase in the length of the four bonds adjacent to the lone-pair because it repels the adjacent *cis* bonding pairs more strongly than the opposite *trans* bonding pair. In BrF_5 the bonds in the base of the square pyramid have a mean length of 1·79 Å, whereas the bond to the apex has a length of 1·68 Å. Other examples are discussed later.

3.3 VARIATION OF BOND ANGLES WITH LIGAND ELECTRONEGATIVITY

We have seen that bonding electron pairs are smaller than non-bonding pairs, taking up less room in the valence shell and repelling other adjacent electron pairs less strongly. Bonding pairs also are expected to vary somewhat in size, depending mainly on the electronegativity of the ligand that they are binding. The greater the electronegativity of the ligand the more it contracts the charge cloud of the bonding pair and attracts it to itself.

The effect of the very electronegative fluorine ligand in decreasing the size of the bonding pair can be shown by calculating the radius of the bonding pair from the lengths of bonds to fluorine using the equation

$$r_e = (d_{X-F} - r_{\text{core X}} - r_{\text{core F}}).$$

The radii of bonding electron pairs in fluorides obtained in this way for some non-metallic elements are given in Table 3.2 and it may be seen that except for nitrogen, oxygen, and fluorine, which are again anomalous, the sizes of the bonding electron pairs in fluorides are consistently less than those listed in Table 2.1, which were obtained from the covalent radii of the elements and may therefore be regarded as average values for all bonds. It may be seen that phosphorus, sulphur, and fluorine can all achieve six electron-pair valence shells in their fluorides which is in accord with the known molecules PF_6, SF_6, and ClF_5, and that argon appears to be limited to four

Table 3.2 Bonding electron-pair radii for fluorides (Å)

	r_{X-F}	Σr_{core}	r_e	r_{core}/r_e	Co-ordination number
C	1·32	0·22	0·55	0·27	4
N	1·37	0·18	0·60	0·19	3–4
O	1·42	0·16	0·63	0·14	3
F	1·44	0·14	0·65	0·11	2
P	1·58	0·41	0·59	0·58	6
S	1·58	0·36	0·61	0·48	6
Cl	1·60	0·33	0·63	0·41	6
Ar	1·60*	0·20	0·65	0·35	4
Se	1·68	0·49	0·59	0·71	8
Br	1·68	0·46	0·61	0·64	7–8
Kr	1·68*	0·43	0·63	0·57	6

*Extrapolated Values.

electron pairs in its valence shell, even when these are decreased in size by bonding to fluorine.

The effect of increasing ligand electronegativity is illustrated in Fig. 3.10. A given electron pair on a given nucleus has its maximum

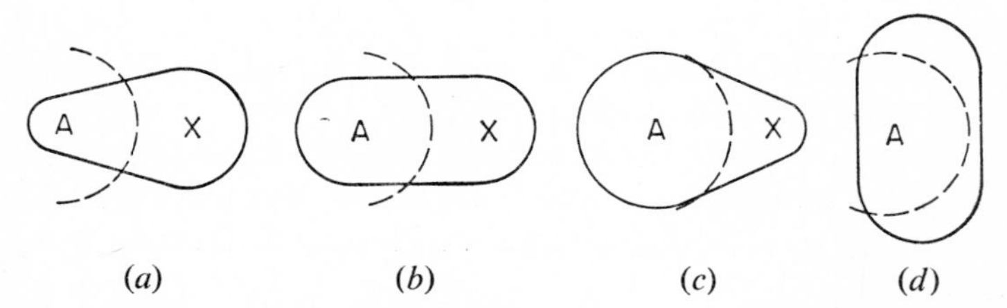

FIG. 3.10 The effect of the electronegativity of a ligand X on the space occupied by the bonding electron pair in the valence shell of a central atom A: (*a*) electronegativity of X greater than that of A; (*b*) electronegativity of X equal to that of A; (*c*) electronegativity of X less than that of A; (*d*) hypothetical case of ligand X of zero electro negativity, i.e., the electron pair is a lone-pair on A.

size and occupies a maximum amount of the surface when it is a lone-pair; it becomes smaller if it is used for bonding and decreases in size as the ligand nucleus attracts it more strongly. Thus, with increasing ligand electronegativity, a bonding pair takes up a decreasing amount of space in the valence shell of the central atom and interacts less strongly with neighbouring electron pairs. Consequently, bond angles decrease with increasing ligand electronegativity as in the following examples:

H_2O 104·5°,	F_2O 103·2°,	NH_3 107·3°,	NF_3 102°
PI_3 102°,	PBr_3 101·5°,	PCl_3 100·3°,	PF_3 97·8°
AsI_3 100·2°,	$AsBr_3$ 99·7°,	$AsCl_3$ 98·7°,	AsF_3 96·0°.

3.4 MULTIPLE BONDS

The shapes of molecules containing double and triple bonds can be readily obtained on the basis of the principles previously outlined, by taking account of the special shapes of the orbitals associated with multiple bonds. Since a double bond consists of two pairs of shared electrons, its shape can be adequately represented by two spheres somewhat squashed together by the attraction of two atomic cores, thus in the first approximation, the overall shape of a double-bond orbital is roughly cylindrical or ellipsoidal rather than spherical as shown in Fig. 3.11. Similarly, the three electron pairs of a triple

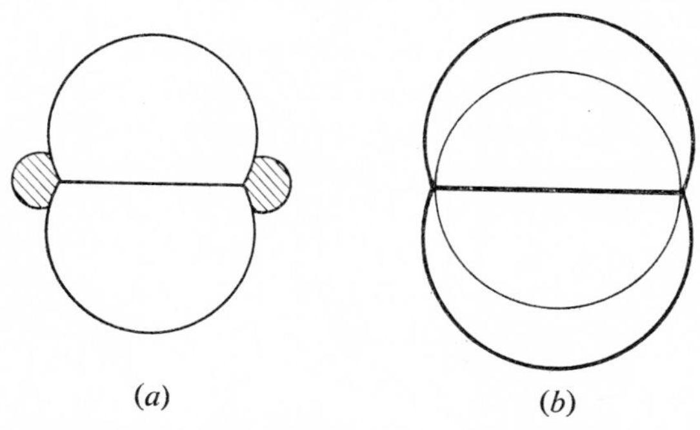

FIG. 3.11 (*a*) approximate ellipsoidal shape of a double bond orbital; (*b*) comparison of the size and shape of a double bond orbital (heavy line) with the approximately spherical shape of a single bond orbital (thin line).

bond can be imagined to be squashed together into the disc or oblate ellipsoidal shape shown in Fig. 3.12. The approximate shapes of molecules containing multiple bonds can then be predicted simply from the total number of orbitals, including lone-pair orbitals, single bond orbitals and multiple bond orbitals, that surround the core of the central atom, a multiple bond orbital for this purpose counting as a single orbital irrespective of the fact that it contains two or even three electron pairs. This is justifiable because the electron pairs of a multiple bond are necessarily held together in the same space and are unable to separate to the maximum possible distance apart as they would if they were single bond pairs. We then apply the rule that two orbitals adopt a linear arrangement, three a planar triangular arrangement, four a tetrahedral arrangement, etc.

Thus for carbon dioxide the two double bond orbitals on the

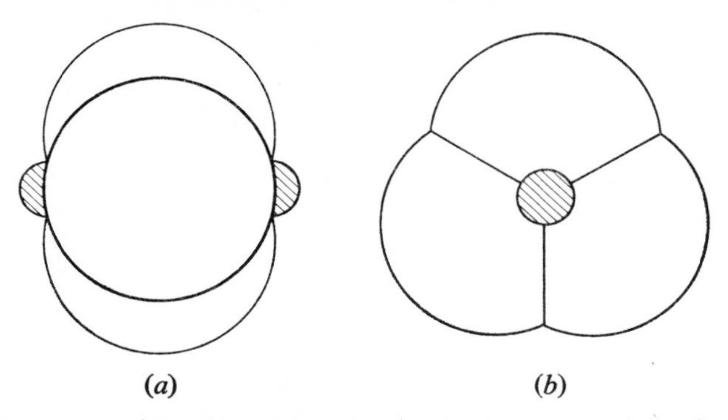

FIG. 3.12 Triple bond orbital: (*a*) side view; (*b*) end-on view.

central atom are expected to adopt a collinear arrangement and carbon dioxide is predicted to be a linear molecule (Fig. 3.13). Moreover the two double bond orbitals will minimize their mutual interaction if they adopt a mutually perpendicular arrangement and thus

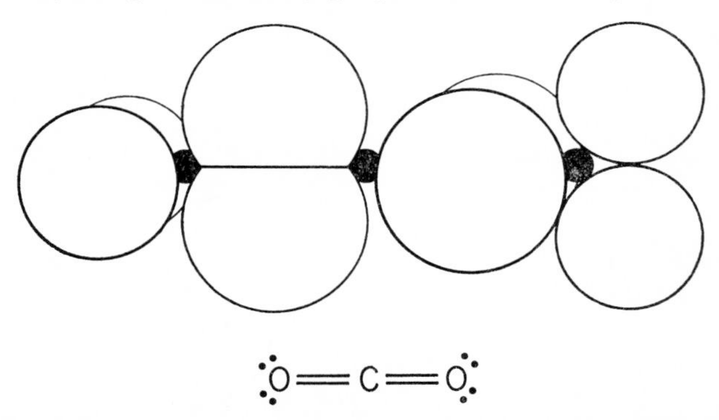

FIG. 3.13 Structure of carbon dioxide based on a collinear arrangement of two double bond orbitals around the central carbon atom.

an approximately tetrahedral arrangement of four electron pairs around the central carbon atom is maintained. Similarly, a single bond and a triple bond adopt a collinear arrangement in molecules such as H—C≡N and H—C≡C—H (Fig. 3.14). Since three orbitals on a central core adopt a planar triangular arrangement, the two single and one double bond on carbon in molecules such as ethylene

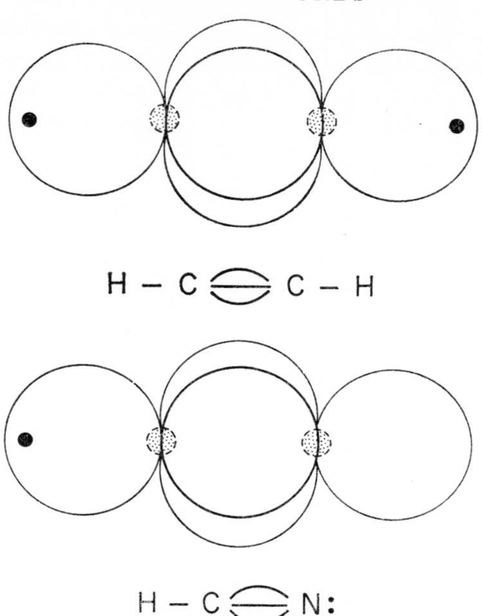

FIG. 3.14 Structures of the linear hydrogen cyanide and acetylene molecules.

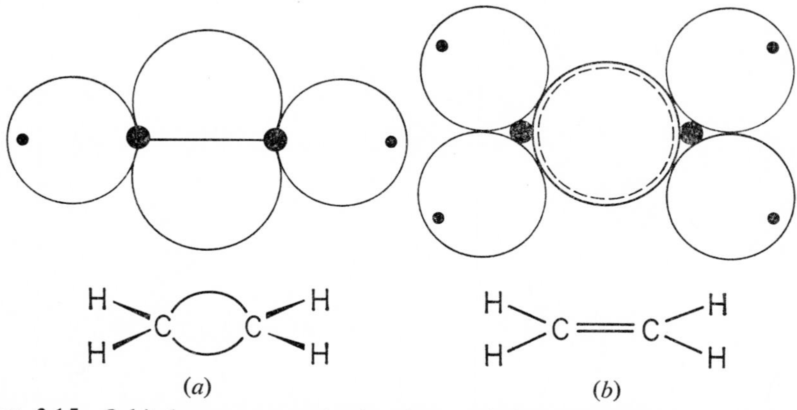

FIG. 3.15 Orbital arrangement in the planar ethylene molecule: (a) side view; (b) top view.

and acetone adopt this coplanar arrangement with bond angles of approximately 120° (Fig. 3.15). Molecules such as SO_2F_2, $SOCl_2$, and F_3SN have structures based on a tetrahedral arrangement of four orbitals around the central, sulphur, two double-bond orbitals and two single-bond orbitals in SO_2F_2, two single-bond orbitals a lone-pair orbital and a double-bond orbital in $SOCl_2$, and three single-bond orbitals and a triple-bond orbital in F_3SN (Fig. 3.16). The shapes of molecules containing multiple bonds and with co-ordination numbers up to six are summarized in Table 3.3.

The general shape of any molecule containing a multiple bond can

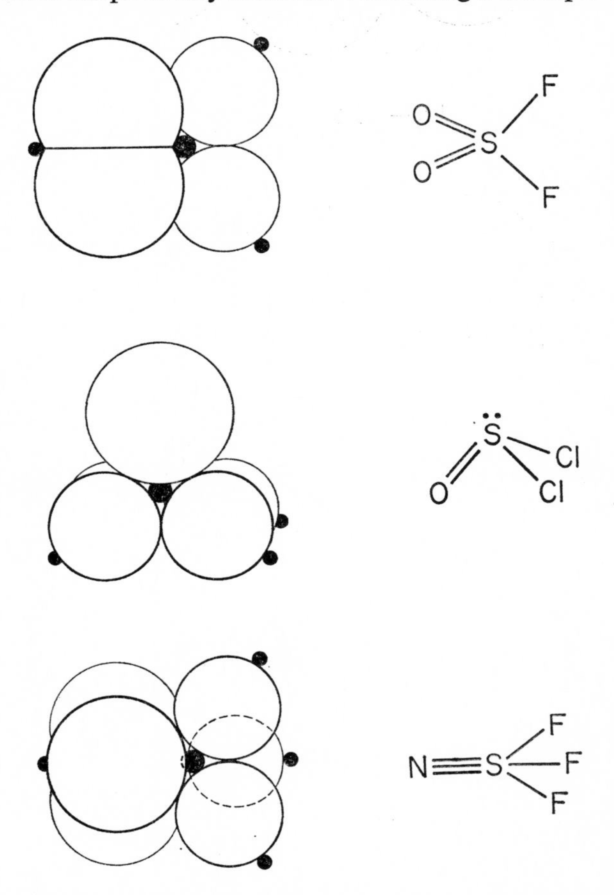

FIG. 3.16 Orbital arrangements in the SO_2F_2, $SOCl_2$, and NSF_3 molecules.

always be correctly predicted by ignoring the difference in size and shape of a double- or a triple-bond orbital and a single-bond orbital. However, in considering the differences in bond angles from the ideal angles it is necessary to take into account the larger size and different shape of the multiple-bond orbitals. Thus we expect that a double-bond orbital will be slightly larger than a single-bond orbital across its short diameter and considerably larger than a single bond orbital but somewhat smaller than two single-bond orbitals along its long diameter.

In the ethylene molecule, or in any molecule containing the $>C=$ bond arrangement as a consequence of the trigonal arrangement of these orbitals around carbon, we expect approximately 120° bond angles. The double-bond orbital in ethylene will have its long axis perpendicular to the plane of the molecule giving an approximately tetrahedral arrangement of the two electron pairs of the double bond and the two electron pairs of the C—H bonds around each carbon atom. In the plane of the molecule, the larger size of the double bond will cause the angle between the two single bonds to be slightly less than 120°. This is generally found to be the case as may be seen from the data given in Table 3.4. Similar considerations apply

Table 3.4 $>C=$bond angles

	$\widehat{XCX}$
$H_2C{=}CH_2$	117·7°
$F_2C{=}CH_2$	109·3
$Cl_2C{=}O$	111·3
$H_2C{=}O$	115·8
$F_2C{=}O$	108·0
$(CH_3)_2C{=}CH_2$	115·3
$Cl_2C{=}CH_2$	114·0

to ketones and in general to any molecule containing the planar bond arrangement $>M=$.

In a tetrahedral arrangement of four orbitals, one of which is a double-bond orbital, the angle involving the double bond is always larger than the tetrahedral angle, leaving the angle between the single bonds less than the tetrahedral angle—as may be seen from the data for $=P\!\!<$ compounds given in Table 3.5. Similarly if there are two double-bond orbitals in a tetrahedral arrangement of four orbitals, the largest angle is always that between the two double bonds, as is illustrated by the data for some sulphuryl compounds that is also given in Table 3.5. We note also from the data given in this table that the XPX bond angle decreases as predicted with increasing ligand electronegativity. If one of the ligands in a

Table 3.5 Bond angles in AX_4 molecules containing double bonds

			$\widehat{XSX}$	$\widehat{OSO}$
POF_3	102·5°	F_2SO_2	96·1°	124·0°
$POCl_3$	103·6	Cl_2SO_2	112·2	119·8
$POBr_3$	108·0	$(NH_2)_2SO_2$	112·1	119·4
PSF_3	100·3	$(CH_3)_2SO_2$	115·0	125·0
$PSCl_3$	100·5			
$PSBr_3$	106·0			

pyramidal AX_3E molecule is bonded by a double bond, then the angles involving this bond will be larger than the other angles. Table 3.6 shows that this is generally the case.

Table 3.6 Bond angle in AX_3E molecules containing a double bond

	XSX	XSO
F_2SO	92·8°	106·8°
Br_2SO	96	108
$(CH_3)_2SO$	100	107
$(C_6H_5)_2SO$	97·3	106·2
$SeOCl_2$	106	114

In many molecules some bonds must be ascribed partial double-bond character, i.e., bond orders between 1 and 2. For example, (1) is a localized electron-pair structure for the carbonate ion. Experi-

$O{=}C(O^-)(O^-)$ **(1)** $\quad$ $^-O{-}C({=}O)(O^-)$ **(2)** $\quad$ $^-O{-}C(O^-)({=}O)$ **(3)**

ment shows, however, that all the bonds are the same length. This is described in valence-bond theory by supposing that there is resonance between (1) and structures (2) and (3). An exactly analogous situation is found in the nitrate ion which may be described in terms of resonance between structures (4), (5), and (6). Each of the bonds may be regarded as having partial double-bond character or a bond order of $1\frac{1}{3}$. Since there are three equivalent bonds in these molecules and

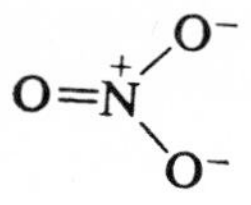

(4) 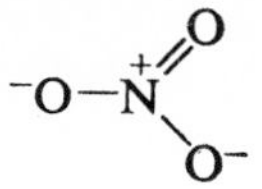 (5) 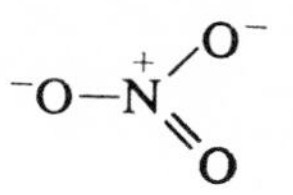 (6)

no non-bonding electrons on the central atom, we predict a trigonal planar arrangement for the three bonds as is in fact observed. They may also be conveniently described by the single formulae (7) and (8) in which the dotted line represents a partial bond. The extent of

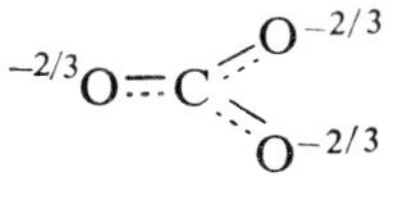

(7) 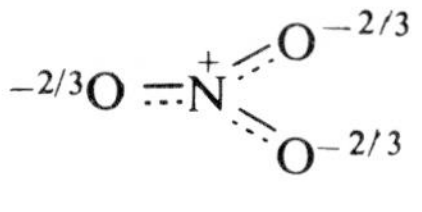 (8)

double bonding, i.e., the bond order, is in fact not particularly relevant to the question of the structures of molecules of this type in which all the bonds are equivalent and there are no lone-pairs on the central atom. Thus BF_3, CO_3^{2-}, NO_3^-, and SO_3 all have a planar trigonal structure based on the expected planar trigonal arrangement of three-bond orbitals around the central atom, irrespective of the bond order which formally at least may be regarded as 1 for BF_3 (9), $1\frac{1}{3}$ for CO_3^{2-} and NO_3^{2-}, and 2 for SO_3 (10). Even if it is considered

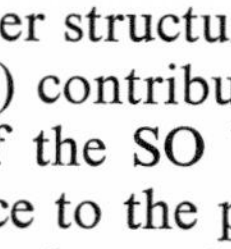

(9) 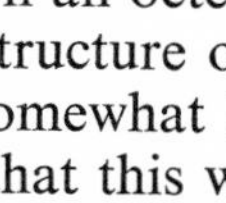 (10) 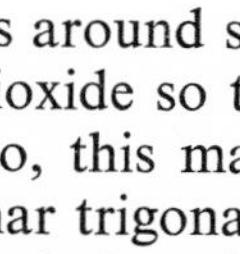 (11)

that other structures based on an octet of electrons around sulphur, e.g., (11) contribute to the structure of sulphur trioxide so that the order of the SO bonds is somewhat less than two, this makes no difference to the prediction that this will be a planar trigonal molecule based on the planar trigonal arrangement of three-bonding orbitals around sulphur.

In terms of the localized electron-pair model that we are using, each bond in CO_3^{2-}, NO_3^- and SO_3 can be imagined to be a double bond, but with the electrons polarized away from the central atom and out of the bonding region and into the non-bonding region of the valence shell of the more electronegative oxygen atom. In each case therefore the central atom is surrounded by three parallel and roughly ellipsoidal double-bond orbitals that are more or less distorted into the non-bonding region of the valence shell of the oxygen to an

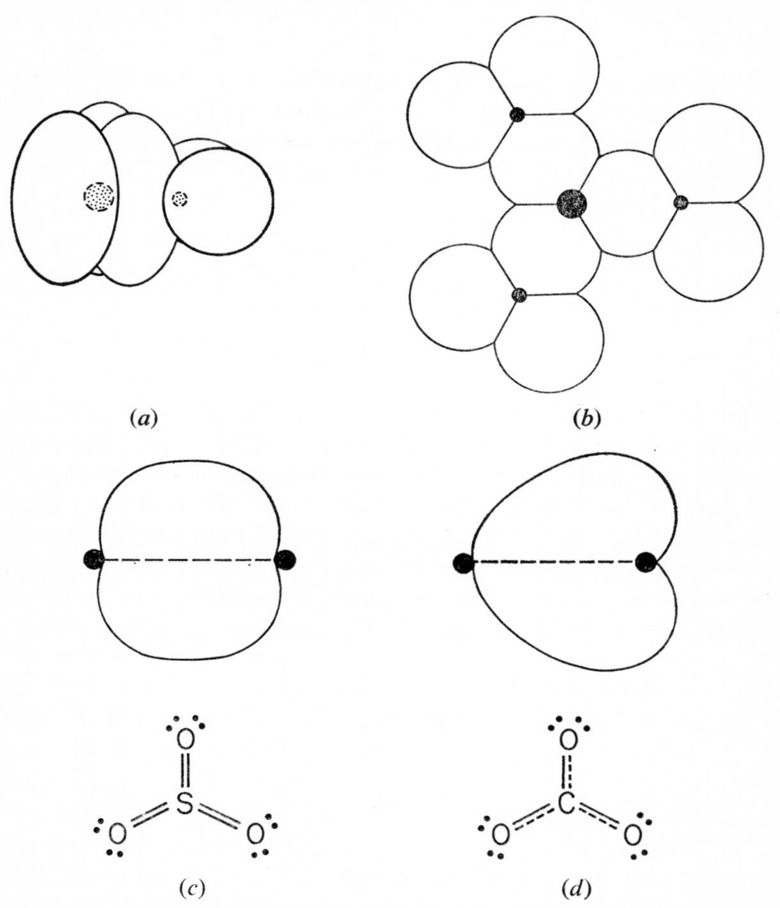

FIG. 3.17 Orbital arrangements in the SO_3 and CO_3^{2-} molecules: (*a*) side view of the double bond orbitals *b* and two of the lone-pair orbitals *l*; (*b*) top view of the double bond and lone-pair orbitals; (*c*) the approximately symmetrical double bond in SO_3; (*d*) the unsymmetrical double bond in CO_3^{2-}.

extent that depends on the relative electronegativities of oxygen and the central atom and on the availability of space for electron density in the valence shell of the central atom. Since sulphur can contain up to six electron pairs in its valence shell when the ligands are electronegative, the bonds in sulphur trioxide may be regarded as being essentially full double bonds with rather symmetrical double-bond orbitals, but the bonds in nitrate and a carbonate presumably have rather asymmetrical double-bond orbitals with considerably less density in the bonding region (Fig. 3.17). Multiple bonds formed by carbon, oxygen, and nitrogen are further discussed in Chapter 6. In the benzene molecule there are three pairs of electrons that cannot be precisely located as is indicated by the two valence-bond structures (12) and (13). The trigonal arrangement of three fixed, single-bond

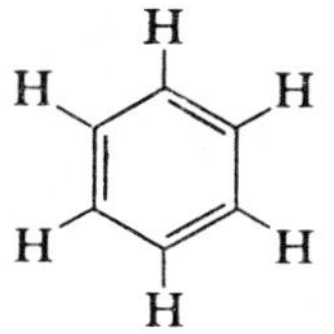

(12) 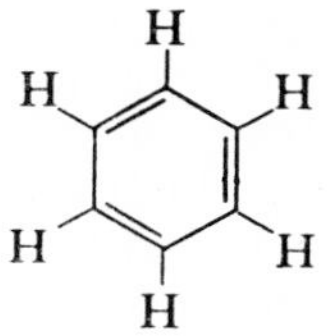 **(13)**

electron pairs around each carbon atom determines the regular hexagonal structure of the benzene molecule, the remaining six electrons are usually described as delocalized: they are assigned to molecular orbitals covering the whole ring and have no direct effect on the geometry. It is not necessary to consider these electrons as being completely delocalized however. The operation of the Pauli principle causes them to stay as far apart as possible so that we might imagine them as forming three pairs and remaining at the maximum distance apart, i.e., at 120° from each other, but the set of three electron pairs being able to circulate freely around the ring of carbon atoms. There is however no reason why these electrons should remain in pairs, as the three electrons of one spin have no correlation with the three electrons of opposite spin, and they are not caused to form into pairs by the attraction of any particular nuclei. Presumably therefore one set is largely independent of the other, and if it is assumed that the electrostatic force between them determines the

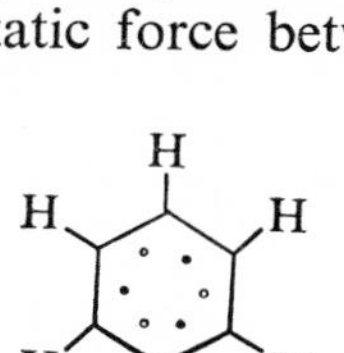

(14)

relative arrangement of the two sets, their resulting staggered arrangement will lead to the presence of one electron in each bonding region and hence to an effective three-electron bond between each carbon atom (14). This is the picture of benzene proposed by Linnett. Again the precise arrangement of the delocalized electrons need not greatly concern us because they have little, if any, effect on the molecular geometry.

3.5 BOND ANGLES AT A CENTRAL ATOM HAVING AN INCOMPLETELY FILLED VALENCE SHELL

In a filled valence shell the electrons occupy all the available space around the central core. For the simple hard-sphere model of an electron pair this would mean that the spheres are closely packed around the central core leaving no room for additional spheres. However, if the valence shell is incompletely filled, then using the hard-sphere model, we may imagine that there is empty space in the valence shell capable of accommodating one or more additional electron pairs, and as the spheres are not touching there is no force resisting the decrease in the angle between any two electron pairs until their orbital spheres touch. In fact, of course, the central core is completely surrounded by the electron density of the valence electrons even if the valence shell is incomplete, i.e., can contain more electrons, but we may think of the electron density as being spread rather thinly, particularly between the bonding electron pairs, so that there is rather little interaction between them. As a consequence, these electron pairs are rather easily pushed together until the distance between them reaches some critical value at which the increasing repulsion between the two charge clouds begins to resist strongly further decrease in the distance between the electron pairs. Thus the elements carbon, nitrogen, oxygen, and fluorine, in the vast majority of their compounds, have valence shells that are completely filled by four electron pairs. These four electron pairs are arranged tetrahedrally, and as they occupy all the available space around the central core the bond angles are not easily distorted from the tetrahedral value. The observed deviations are in no case more than a few degrees (Table 3.7).

Elements in the third and subsequent rows of the periodic table are, however, larger, and have more space in their valence shells. For example, as we saw in Table 2.1 the elements silicon, phosphorus, and sulphur can accommodate a maximum of six electron pairs in

Table 3.7 Bond angles for atoms having a filled valence shell of four electron pairs

H_2O	104·5 ± 0·1°	
NH_3	107·3 ± 0·2	
OF_2	103·2 ± 1	
Cl_2O	110·8 ± 1	
$(CH_3)_2O$	111 ± 3	
$(CH_3)_3N$	108 ± 4	
CH_3Cl	110 ± 0·5	(HCH)
CH_2Cl_2	112·0 ± 0·3	(HCH)
	111·8 ± 0·3	(ClCCl)
$CHCl_3$	110·4 ± 1	(ClCCl)
CH_2F_2	108·3 ± 1	(FCF)
	111·9 ± 4	(HCH)
CH_3NH_2	109·5 ± 1	(HCH)

their valence shells. Since six electron pairs in a valence shell of this size have an octahedral arrangement we assume that the electron pairs do not interact strongly until they are at an angle of approximately 90° to each other. When there are only four electron pairs in such a valence shell they adopt a tetrahedral arrangement because of their mutual interaction, and they do not take up all the space in the

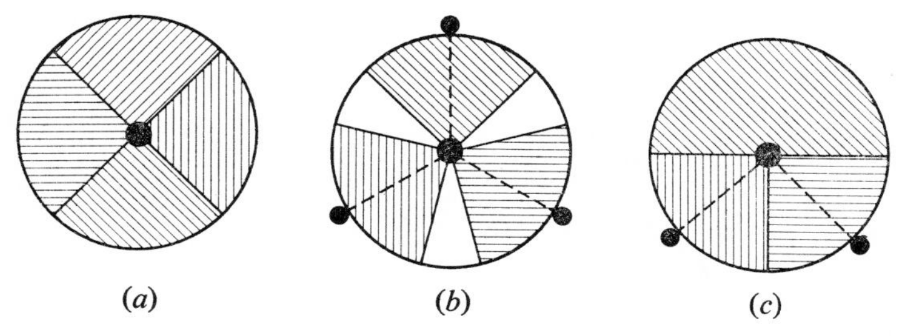

FIG. 3.18 Electron pairs in a 2-dimensional valence shell: (*a*) 4 equivalent electron pairs; (*b*) 3 equivalent bonding pairs leaving 'empty space' in the valence shell; (*c*) 2 bonding pairs and one non-bonding pair which spreads out and forces the bonding pairs together.

valence shell. Hence the bond angles between these four electron pairs are easily distorted. Thus, although in SiH_4 the bond angle must be the tetrahedral angle for reasons of symmetry, when, as in PH_3, one of the four pairs is a non-bonding pair which tends to spread out to take up as much space as possible around the central core, the angle between the three bonding pairs is decreased considerably from the tetrahedral angle towards the limiting angle of 90°. Hence the bond angle in PH_3 is 93·3°. In the H_2S molecule the presence of two non-bonding pairs is expected to cause a further reduction in the bond angle, but as the limiting bond angle is being approached only a small further reduction of the angle to 92·2° occurs. This effect can be illustrated very simply by means of a hypothetical two-dimensional valence shell which is large enough to accommodate four electron pairs (Fig. 3.18). These will make an angle of 90° to each other as shown in (a). If the valence shell contains only three equivalent bonding electron pairs then they will make angles of 120° to each other leaving empty space between the orbitals. If one of the three orbitals is a lone-pair however it will spread out and push the bonding pairs together until they make an angle of approximately 90° to each other. Thus we see that the bond angle is easily distorted from 120° towards the limiting value of 90°.

3.6 BOND ANGLES IN HYDRIDES

Bond angles in hydrides are generally smaller than would be expected from the electronegativity of hydrogen. For example, the bond angles in PCl_3 and PF_3 which are 100·3° and 97·8° respectively, are larger than in PH_3, which has a bond angle of 93·3°, and similarly the bond angles in $AsCl_3$ and AsF_3, which are 98·7° and 96·0° respectively, are larger than in AsH_3, which has a bond angle of 91·8°. In fact, bond angles involving hydrogen are, in general, anomalously small. This can be attributed to the unique nature of hydrogen as a ligand in that it has no other electrons in its valence shell other than the bonding electron pair. This means that in addition to occupying the bonding region this electron pair is also spread out around the hydrogen nucleus, whereas in other molecules which have valence shells which can contain up to at least four electron pairs these other electron pairs occupy much of the valence shell and the bonding electron pair is indeed confined largely to the bonding region. Thus we may conclude that in general the electron density in the bonding region of an A—H bond is less than would be expected from a consideration of the electronegativity of hydrogen alone, and consequently bond angles involving A—H bonds are smaller than would be expected on this basis.

3.7 MULTICENTRE BONDS

There is no reason why an electron pair should be confined to binding two atoms as in the ordinary two-centre covalent bond, and indeed it is possible for a single electron pair to bond three or even more atoms together. Thus, if three atomic cores are attracted by a single electron pair they adopt a triangular arrangement around the electron pair, the electron pair forming what is known as a three-centre orbital. Similarly, four atomic cores would cluster around an

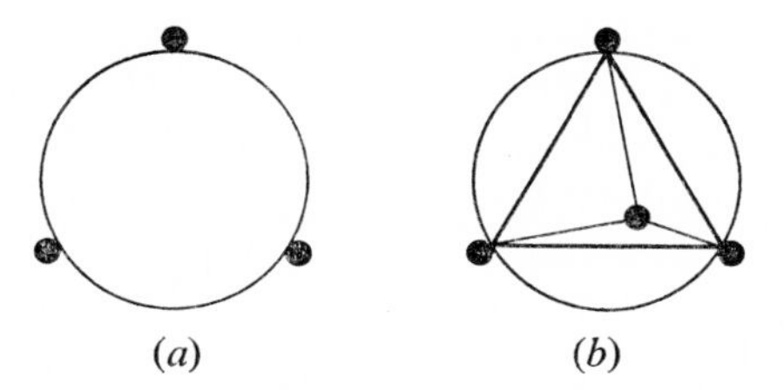

FIG. 3.19 (*a*) Triangular arrangement of 3 nuclei around a single electron pair orbital; (*b*) Tetrahedral arrangement of four nuclei around a single electron pair orbital.

electron-pair orbital in a tetrahedral arrangement to give a four-centre bond (Fig. 3.19). Such multicentre bonds occur typically in electron-deficient molecules in which there are not enough electrons to occupy two-centre orbitals between all the atoms that are bonded together; in such a case multicentre orbitals bond some of the nuclei together. Diborane is a simple example of an electron-deficient molecule which contains two three-centre bonds (Fig. 3.20). Since the pair of electrons in a three-centre bond is shared between three rather than two positive cores the electron density in the valence shell of any one of the atoms bonded together is necessarily less than

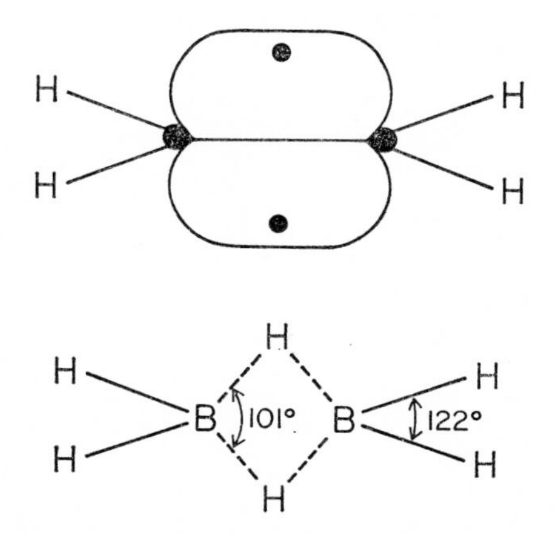

FIG. 3.20 Structure of B_2H_6 showing the two three-centre orbitals.

it would be in a two-centre bond. Thus, in diborane, the angle at boron between the two bridging hydrogens is less than that between the two terminal hydrogens because the bridging hydrogens are bound by three-centre bonds in which the electron density between the hydrogen and boron is only one-half as great as between boron and hydrogen in the terminal bonds.

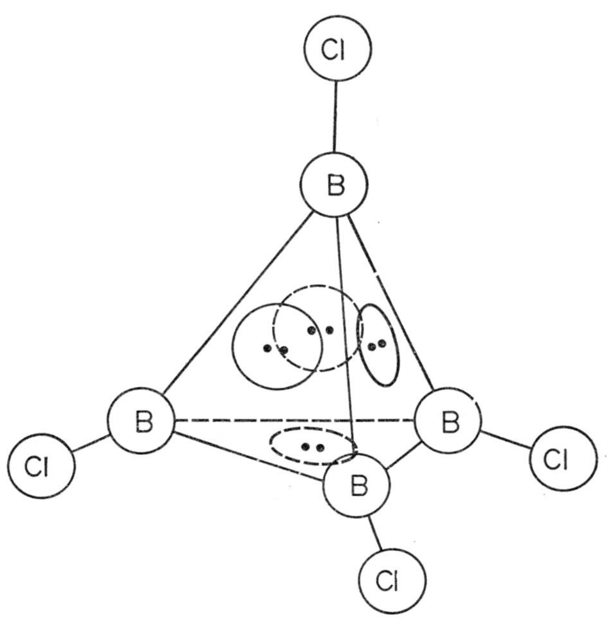

FIG. 3.21 Structure of B_4Cl_4 showing the four three-centre orbitals.

The molecule B_4Cl_4 has the structure as shown in Fig. 3.21 in which there is a tetrahedron of borons with a chlorine bonded to each corner. Assuming that each boron–chlorine bond is a normal, single electron-pair bond, only eight electrons remain to bond the four borons together in the tetrahedron. It appears that in this molecule there are four three-centre bonds between the borons, one for each face of the tetrahedron as shown in Fig. 3.21.

REFERENCES AND SUGGESTIONS FOR FURTHER READING

H. A. BENT, *J. Chem. Ed.*, **45,** 768 (1968).
R. J. GILLESPIE, *J. Chem. Ed.*, **40,** 295 (1963); **47,** 18 (1970).
R. J. GILLESPIE, *Angew. Chem. Internat. Ed.*, **6,** 819 (1967).
J. W. LINNETT, *The Electronic Structure of Molecules*, Methuen, London (1964).
L. PAULING, *Nature of the Chemical Bond*, 3rd Ed., Cornell University Press, 1960.
L. E. SUTTON (Ed.), *Interatomic Distances*, Chemical Society Publications No. 11 (1958), No. 18 (1965).

4

Valence Shells Containing Five Electron Pairs: AX_5, AX_4E, AX_3E_2, and AX_2E_3 Molecules

4.1 ARRANGEMENTS OF FIVE ELECTRON PAIRS

Molecules in which the valence shell of the central atom has five electron pairs show a number of features of special interest that merit a separate discussion. It was shown in Chapter 1 that the favoured arrangements for four, five, and six mutually repelling points (electron pairs) on the surface of a sphere are the tetrahedron, trigonal bipyramid, and octahedron respectively. In the tetrahedron and the octahedron all the vertices are equivalent, whereas in the trigonal bipyramid they are not. In the tetrahedron and the octahedron each vertex has the same number of nearest neighbours in the same directions and at the same distance, while in the trigonal bipyramid the apical vertex has three nearest neighbours at 90° and an equatorial vertex has only two nearest neighbours at 90°. We have so far, for simplicity, assumed that the mutually repelling points (electron pairs) are situated on the surface of a sphere. We may now remove this assumption and assume that the distances of the electron pairs from the central core are determined by the balance of their mutual repulsions and their attraction by the central core. For the tetrahedron and octahedron, since all the electron pairs are equivalent even if the restriction of confining them to some given spherical surface is removed, they will nevertheless remain on the surface of a sphere. However, this is not the case for the trigonal bipyramidal arrangement. It can be shown that for a force law of the type $F = 1/r^n$ the electron pairs only remain on the surface of a sphere if $n = 3{\cdot}4$. For $n > 3{\cdot}4$, if the electron pairs are initially placed on the surface of a sphere, the force on the axial pairs is greater than on the equatorial pairs, and if the electron pairs are no longer confined to a sphere but allowed to take up their true equilibrium positions the axial pairs will be at a greater distance from the nucleus than the equatorial pairs. On the other hand, if $n < 3{\cdot}4$, then at equilibrium the equatorial pairs will be at a greater distance from the nucleus than the axial pairs. Since, as was discussed in Chapter 2, n has a fairly large value that is certainly greater than 3·4 we may predict that equatorial electron pairs will be closer to the central core than axial electron pairs in a trigonal bipyramid arrangement. The same conclusion may be reached by considering the hard-sphere model. If we allow each sphere to touch the central core then the precise arrangement is indeterminate (Fig. 4.1). In a trigonal bipyramid arrangement the equatorial spheres do not touch each other and are further

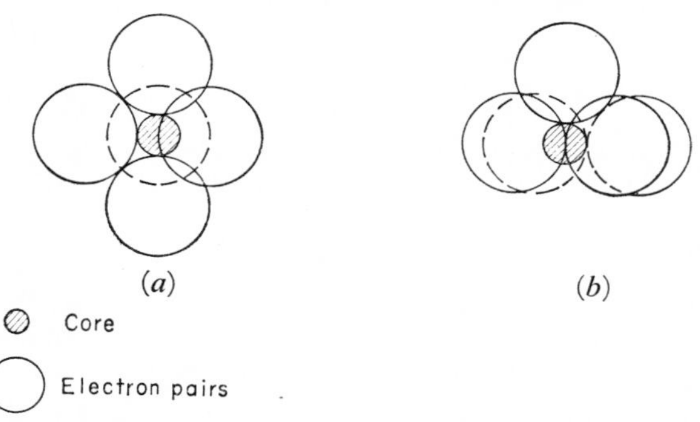

FIG. 4.1 (*a*) Trigonal bipyramid arrangement of five electron pairs around a central core. The three equatorial pairs do not touch each other. (*b*) Square pyramidal arrangement of five electron pairs around a central core. All five electron pairs touch each other. Any arrangement of the five electron pairs intermediate between (*a*) and (*b*) is also possible.

from each other than from the axial pairs. They can only be made to touch each other by reducing the angle between the equatorial pairs to 90°, which then gives a square pyramid arrangement. Thus, in terms of this model, any equatorial bond angle between 90° and 120° is possible, i.e., any structure is possible—from the square pyramid to the trigonal bipyramid. The hard-sphere model corresponds to $n = \infty$, but for any value of n less than infinity the trigonal bipyramid becomes at least slightly more stable than the square pyramid. If we remove the restriction that all the electron-pair spheres are at the same distance from the core, but instead require that all neighbouring spheres should touch and that all distances between neighbouring electron pairs are the same, the bipyramid then consists of two tetrahedra sharing a face, and the distance of the axial electron pairs from the central core is $\sqrt{2}$ times that of the equatorial pairs (Fig. 4.2). It seems probable that most real molecules will adopt an

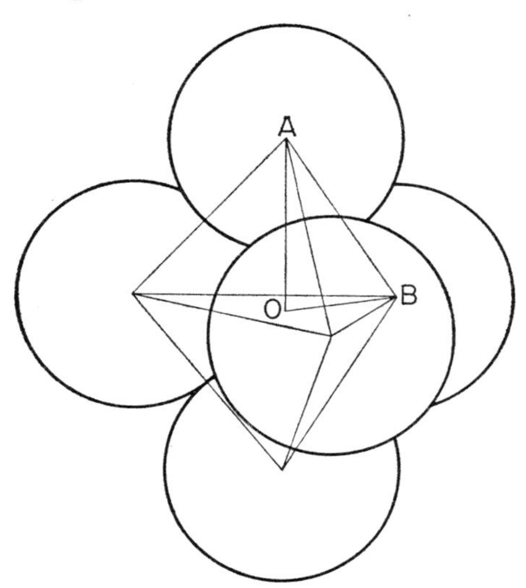

FIG. 4.2 Trigonal bipyramid arrangement of five electron pairs around a central core in which all electron pairs are equidistant from each other. The trigonal bipyramid in this case consists of two tetrahedra sharing a face. The axial electron pairs are at a greater distance from the core than the equatorial electron pairs; $OA = \sqrt{2}OB$.

intermediate arrangement of five electron pairs in which the equatorial pairs are somewhat closer to the axial pairs than they are to each other and in which the ratio of the distance of the axial pairs from the central core is somewhat less than $\sqrt{2}$.

4.2 PROPERTIES OF MOLECULES WITH FIVE-ELECTRON-PAIR VALENCE SHELLS

The above discussion leads us to a number of predictions concerning molecules that have five electron pairs in their valence shells.

1. AX_5 molecules with a spherical central core will have a trigonal bipyramid shape and AX_4E, AX_3E_2, and AX_2E_3 molecules will have shapes based on a trigonal bipyramid arrangement of five electron pairs. A number of AX_5 molecules in which A has a spherical core have had their structures determined, and with two exceptions they are all trigonal bipyramids. The exceptions are antimony pentaphenyl and $InCl_5^{2-}$ which have square pyramid structures in the

Table 4.1 Equatorial and axial radii for AX_5, AX_4E, and AX_3E_2 molecules*

	r_{eq}(Å)	r_{ax}(Å)	r_{ax}/r_{eq}
	AX_5		
PCl_5	1·05	1·20	1·14
PF_5	0·89	0·94	1·06
$P(C_6H_5)_5$	1·08	1·22	1·13
CH_3PF_4	0·90	0·97	1·08
$(CH_3)_2PF_3$	0·91	1·00	1·10
Cl_2PF_3	0·95	1·06	1·12
$SbCl_5$	1·32	1·44	1·09
$(C_2H_2Cl)_3SbCl_2$	1·38	1·46	1·06
$(CH_3)_3SbCl_2$	1·3	1·50	1·15
$(CH_3)_3SbBr_2$	1·3	1·49	1·15
$(CH_3)_3SbI_2$	1·3	1·55	1·19
$(C_6H_5)_3BiCl_2$	1·47	1·61	1·10
$(CH_3)_2SnCl_3^-$	1·42, 1·36	1·55	1·11
	AX_4E		
SF_4	0·91	1·01	1·11
OSF_4	0·90	0·96	1·07
$(C_6H_5)_2SeBr_2$	1·14	1·36	1·19
$(C_6H_5)_2SeCl_2$	1·14	1·31	1·15
$(CH_3C_6H_4)_2SeCl_2$	1·17	1·40	1·20
$(CH_3C_6H_4)SeBr_2$	1·17	1·40	1·20
$(CH_3)_2TeCl_2$	1·33	1·52	1·14
$(C_6H_5)_2TeBr_2$	1·37	1·54	1·12
$O_2IF_2^-$	1·27	1·36	1·07
	AX_3E_2		
ClF_3	0·94	1·04	1·11
$C_6H_5ICl_2$	1·23	1·46	1·19
BrF_3	1·08	1·18	1·09
	AX_2E_3		
XeF_2	—	1·35	
ICl_2^-	—	1·58	
I_3^-	—	1·57	

* Where it is necessary to distinguish between axial and equatorial ligands the equatorial ligands are written in the formula before the central atom.

crystal. We have already noted that for $n = \infty$ the square pyramid structure is equally as probable as the trigonal bipyramid structure, and for $n < \infty$ it is of only slightly higher energy, thus it is not surprising that in some cases other factors lead to the square pyramid structure being preferred to the trigonal bipyramid. Moreover we might expect that for heavier atoms with larger valency shells in which the interaction between axial and equatorial electron pairs is relatively small the difference in energy between the trigonal bipyramid and the square pyramid arrangement may become negligibly small. Thus it is interesting to note that the square pyramid structure has been observed for the relatively heavy elements In and Sb. Moreover although antimony pentaphenyl has the square pyramid structure both arsenic and phosphorus pentaphenyl have the expected trigonal bipyramid structure. Presumably the difference in energy between the square pyramid and the trigonal bipyramid structures for antimony pentaphenyl are small enough that packing considerations predominate in the crystal. There is at present no evidence on the structure of antimony pentaphenyl in solution. AX_4E, AX_3E_2, and AX_2E_3 molecules have the shapes shown in Fig. 1.5 (p. 12) which are all based on a trigonal bipyramid arrangement of five electron pairs in the valence shell of A. Examples of each of these types of molecules are given in Table 4.1.

2. As non-bonding or lone-pairs of electrons are larger than bonding electron pairs and exert a greater force on neighbouring electron pairs, then such non-bonding pairs will occupy the equatorial positions of a trigonal bipyramid arrangement because there is more room for them in these positions and because their interactions with neighbouring electron pairs are thereby minimized. This may be shown very simply by means of the 'spheres and elastic band' model described in Chapter 2 (Fig. 4.3).

3. In general, we expect to find the axial electron pairs at a greater distance from the nucleus than the equatorial pairs. Hence, using the same covalent radius for the ligand in both the axial and equatorial positions, the covalent radius of the central atom should be greater in the axial than in the equatorial direction. In all molecules of this type that have been studied this has proved to be the case, as may be seen in Table 4.1. The ratio of the axial and equatorial covalent radii r_{ax}/r_{eq} falls in the range 1·06–1·20, which is, as expected, somewhat less than the value of $\sqrt{2}$ predicted by the hard-sphere model, with all adjacent spheres touching.

4. The lone-pairs in the equatorial positions, because they are larger than the bonding pairs, will cause the bond angles to be

FIG. 4.3 Photograph of a model showing the preference of the larger unshared pairs for the equatorial positions of a trigonal bipyramid.

smaller than the ideal angles, axial–equatorial = 90°, equatorial–equatorial = 120° and axial–axial = 180°. This is generally found to be the case as shown in Table 4.2. The three possible exceptions are discussed in Chapter 7.

Table 4.2 Bond angles in AX_4E and AX_3E_2 molecules

AX_4E	equatorial–equatorial	axial–axial
SF_4	101°	173°
$(CH_3)_2TeCl_2$	98·2	172·3
$IO_2F_2^-$	100	180
$(C_6H_5)_2SeBr_2$	110	180
$(p\text{—}CH_3.C_6H_4)_2SeBr_2$	108	183
$(p\text{—}CH_3.C_6H_4)_2SeCl_2$	106·5	182
$(C_6H_5)_2TeBr_2$	94·2	182

AX_3E_2	axial–equatorial
ClF_3	87·5°
BrF_3	86·2
$C_6H_5ICl_2$	86

5. Because of their large size, multiple bonds are predicted to occupy the equatorial positions of a trigonal bipyramid. This is known to be the case in $IO_2F_2^-$ and SOF_4 (Fig. 4.4). The recently prepared XeO_2F_2 molecule is predicted to have the lone-pair and

two doubly-bonded oxygen atoms in the equatorial positions of a trigonal bipyramid giving the structure shown in Fig. 4..4 The molecule XeO_3F_2, which has also recently been prepared, is predicted to have a trigonal bipyramid structure with all three oxygen atoms in equatorial positions. In a molecule containing one or two double bonds we may expect a distortion of the ideal bond angles quite similar to that produced by one or two lone-pairs in the equatorial

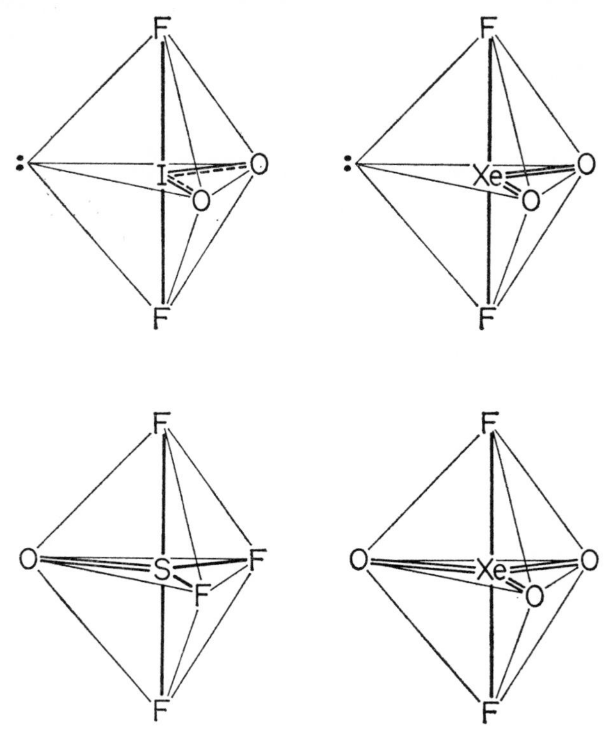

FIG. 4.4 Structures of $IO_2F_2^-$, XeO_2F_2, SOF_4 and XeO_3F_2 all of which are based on a trigonal bipyramid arrangement of five orbitals.

positions. Only for the SOF_4 molecule is data available to check this prediction, and in this case it has been found that all four fluorine atoms are bent away from the oxygen atom as predicted.

6. If there is more than one type of ligand, the least electronegative ligands, i.e., those with the largest bonding orbitals, will occupy the equatorial positions and the most electronegative will occupy the axial positions. For example, in PCl_3F_2 and PCl_2F_3 the chlorine atoms occupy the equatorial positions, in CH_3PF_4 and $(CH_3)_2PF_3$

the methyl groups occupy the equatorial positions, and in $(CH_3)_3SbCl_2$ the methyl groups are again in the equatorial positions. Indeed, in all molecules that have been studied containing alkyl groups and halogens, the more electronegative halogens occupy the axial positions.

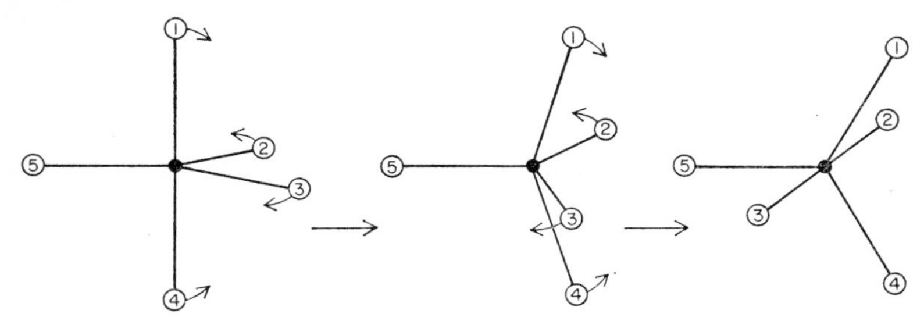

FIG. 4.5 Intramolecular ligand exchange (pseudo-rotation) in trigonal bypyramidal molecules.

4.3 INTRAMOLECULAR LIGAND EXCHANGE (PSEUDO-ROTATION)

Intramolecular exchange of ligands occurs very readily in trigonal bipyramid molecules. For five hard orbitals we have seen that the best arrangement is indeterminate, and no energy is required to transform the trigonal bipyramid to the square pyramid. The trigonal bipyramid can then be reformed in such a way as to exchange the original equatorial and axial positions (Fig. 4.5). This process has been called pseudo-rotation, as the result of the intramolecular exchange of ligands for five identical ligands is the same as if the molecule had been rotated through 90° about the axis through the central atom and ligand 5. For soft orbitals the square pyramid has only a slightly higher energy than the trigonal bipyramid and it represents the transition state between the two trigonal bipyramidal forms. Hence the activation energy for the intramolecular exchange of ligands is small and exchange occurs rapidly. Thus the ^{19}F n.m.r. spectrum of PF_5 is a single doublet due to P–F spin–spin coupling, although because of the stereochemical non-equivalence of the axial and equatorial fluorines two doublets would in fact be expected of relative areas 2 : 3 and which would have further fine structure due to fluorine coupling. As the P–F spin–spin coupling is retained despite the collapse of the expected signals from axial and equatorial fluorines it is concluded that there is rapid intramolecular exchange of fluorines as shown in Fig. 4.5.

This intramolecular exchange is a very facile process because the trigonal bipyramid and square pyramid geometries have very similar energies and the activation energy for the process is therefore low. This activation energy is low because at no time during the exchange do any of the electron pairs come appreciably closer than 90°, which is the critical angle for phosphorus. In contrast, it may be noted that a similar internal exchange in an octahedral molecule, for example the conversion of a *cis* isomer to a *trans* isomer, would involve a transition state with angles between electron pairs considerably smaller than 90°. Strong orbital repulsions prevent such an internal exchange in this case.

Apparent magnetic equivalence of the ligands in a trigonal bipyramid structure has also been observed for $Fe(CO)_5$ and SF_4, although in the latter case the intramolecular exchange can be slowed sufficiently at low temperatures so that separate signals from axial and equatorial fluorines can be observed. This is an interesting observation and implies that the activation energy for the intramolecular exchange is larger in SF_4 than in PF_5; perhaps the larger lone-pair hinders the required vibrational motion of the fluorines.

REFERENCES AND SUGGESTIONS FOR FURTHER READING

S. Berry, *J. Chem. Phys.*, **32,** 933 (1960).

R. J. Gillespie, *J. Chem. Soc.*, 4672 (1963); 4678 (1963).

E. L. Muetterties and R. A. Schunn, *Quart. Rev. Chem. Soc.*, **20,** 245 (1966).

L. E. Sutton (Ed.), *Interatomic Distances*, Chemical Society Special Publication No. 11 (1958); No. 18 (1965).

5

Valence Shells Containing More Than Six Electron Pairs

5.1 ARRANGEMENTS OF SEVEN TO TWELVE ELECTRON PAIRS IN A VALENCE SHELL

For the purposes of simplicity this chapter is largely confined to a consideration of discrete molecules of the type AX_n, where X is a unidentate ligand, and no attempt is made to give a comprehensive discussion of all molecules exhibiting higher co-ordination numbers. In fact, rather few simple complexes of the type AX_n are known, as higher co-ordination numbers are often stabilized by chelating ligands or in polymeric structures. With increasing co-ordination numbers the differences between different geometries become progressively smaller, and if the ligands are not all equivalent, or if there are other reasons why the idealized geometry might be distorted, e.g., packing considerations in the solid state, then it becomes difficult to distinguish the distorted form of one idealized polyhedron from the distorted form of another idealized polyhedron. Again, because the energies of the different polyhedra for a given co-ordination number are very similar, and the activation energies for the transformation of one form to another are expected to be rather small, rapid conversion of one form to the other may occur in the liquid and gaseous states, and some physical methods, e.g., nuclear magnetic resonance may be unable to distinguish between the various possible structures.

Only in the case of the seven-electron-pair valence shell are there any examples of a valence shell containing an unshared pair. For valence shells containing eight or more electron pairs all the electron pairs are bonding pairs in the known examples, and therefore one cannot distinguish between ligand–ligand and bond–bond repulsions as the important factor in determining the structure. It seems unlikely that many examples will be found of large valence shells containing lone-pairs, because such high co-ordination numbers are expected mainly for transition metals and lanthanides and actinides. In these latter cases, any unshared electrons occupy inner *d* and *f* orbitals where they have a smaller effect on stereochemistry than an unshared pair in a valence shell (Chapter 8).

Maximizing the least distance between points on a sphere leads to

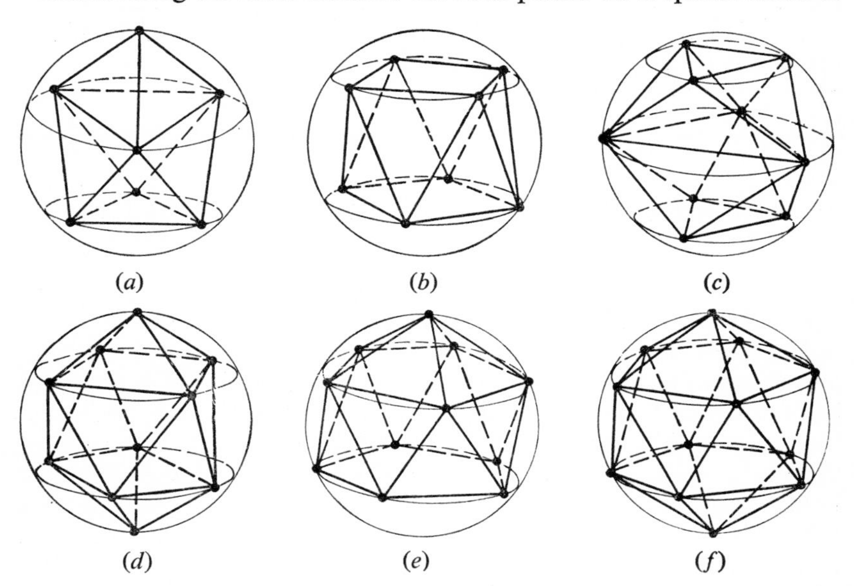

FIG. 5.1 Arrangements of 7 to 12 points on a sphere that maximize their distance apart. (*a*) 7 points—monocapped octahedron; (*b*) 8 points—square antiprism; (*c*) 9 points—tricapped trigonal prism; (*d*) 10 points—bicapped square antiprism; (*e*) 11 points—monocapped pentagonal antiprism; (*f*) 12 points—icosahedron.

the arrangements for seven to twelve points given in Fig. 5.1 and Table 5.1. These are therefore also the expected arrangements for seven to twelve electron pairs according to the hard-sphere model. For eight, nine, and twelve electron pairs replacing hard orbitals ($n = \infty$), where the repulsion between electron pairs is given by ($F = 1/r^n$) by soft orbitals ($n < \infty$) does not lead to any change in the predicted arrangement, but in the other cases other arrangements become stable as n is varied, and these are discussed separately below. For high co-ordination numbers in general, several structures will have similar energies, and for any given value of n a variety of factors, such as the presence of non-equivalent ligands, chelating ligands, and lone-pair electrons, could cause one of these structures to be more stable for a particular molecule than the structures given above, even in the cases of eight, nine, and twelve electron pairs. It

Table 5.1

Number of electron pairs	Geometry
Seven	monocapped octahedron
Eight	square antiprism
Nine	tricapped trigonal prism
Ten	bicapped square antiprism
Eleven	monocapped pentagonal antiprism
Twelve	icosahedron

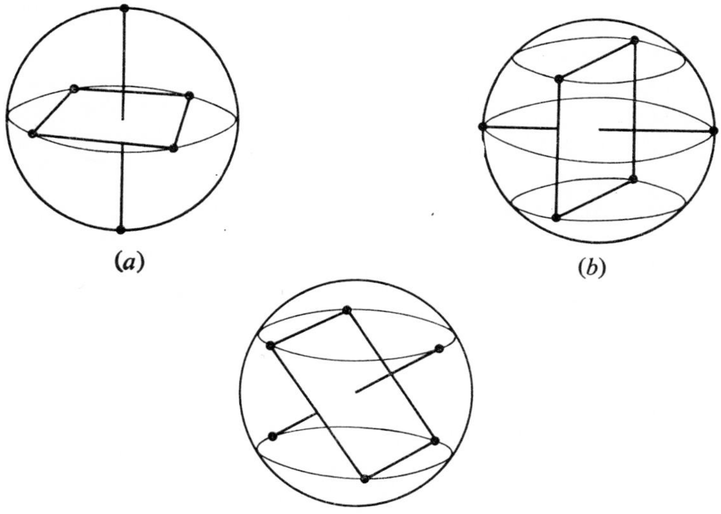

(*a*) (*b*) (*c*)

FIG. 5.2 Arrangements of six electron pairs: (*a*) 1 : 4 : 1; (*b*) 2 : 2 : 2; (*c*) 3 : 3; All three arrangements are identical and are octahedral.

appears that all the probable alternative structures may be derived by means of the following rules concerning the placing of mutually repelling points on a sphere, where each point represents an electron pair.

1. There may be points at one or both poles.
2. There are points located on circles lying between the poles, the plane through each circle being perpendicular to the polar axis.

3. The number of points on successive circles passing from the equator to the pole either stays constant or decreases by two.
4. The points on any given circle maximize their distance apart.
5. The points on successive circles adopt a staggered configuration in the most stable arrangements.

Using these rules we find the following arrangements for six electron pairs: 1 : 4 : 1, 2 : 2 : 2, and 3 : 3, which are all identical with the octahedron, confirming that the octahedron is a uniquely stable arrangement for six electron pairs (Fig. 5.2).

5.2 SEVEN ELECTRON PAIRS

For seven electron pairs the above rules lead to the prediction of the 1 : 3 : 3 arrangement or monocapped octahedron, the 1 : 4 : 2 arrangement or monocapped trigonal prism, the 1 : 5 : 1 arrangement or pentagonal bipyramid, and the 1 : 2 : 2 : 2 arrangement (Fig. 5.3).

It has been shown that as n decreases from infinity the most stable arrangement for seven particles on a sphere changes from the 1 : 3 : 3 monocapped octahedron to the 1 : 2 : 2 : 2 arrangement for intermediate values of n to the 1 : 5 : 1 pentagonal bipyramid for low values of n, but the energies of all three structures are quite similar, and it does not seem possible to predict which structure will be adopted in any particular case. Rather few structures have been determined for seven co-ordinated molecules, and Table 5.2 lists the structures for seven equivalent ligands that have been determined with reasonable certainty. The 1 : 2 : 2 : 2 structure has not been observed, but the rather similar 1 : 4 : 2 monocapped trigonal prism is known in several cases. As in the case of the trigonal bipyramid, because of the non-equivalence of the seven positions in each of these structures, we would not expect all the bond lengths to be the

Table 5.2

1:3:3	1:4:2	1:5:1
$NbOF_6^{2-}$	NbF_7^{2-}	ZrF_7^{2-}
	TaF_7^{2-}	IF_7
		$UO_2F_5^{3-}$
		UF_7^{3-}

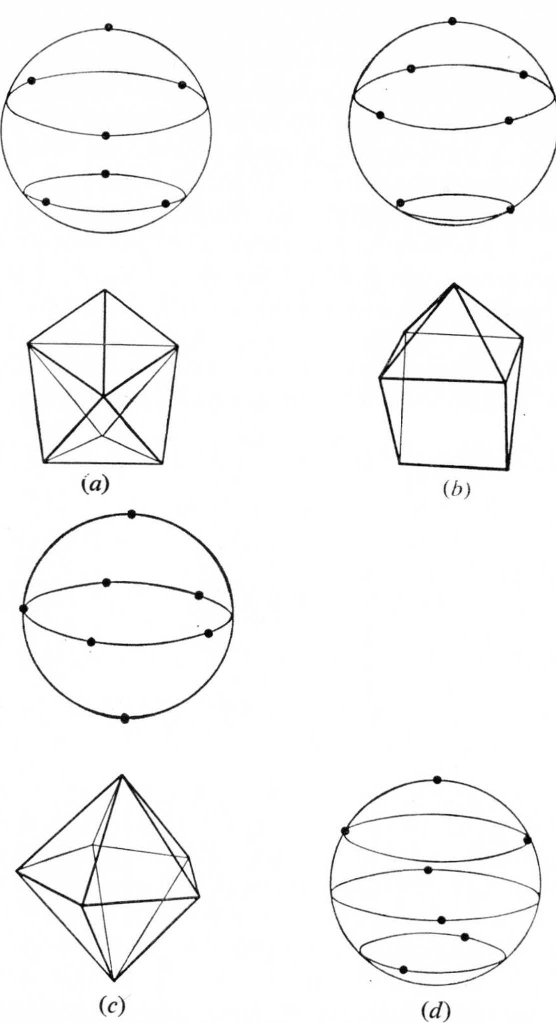

FIG. 5.3 Possible arrangements of seven electron pairs: (*a*) monocapped octahedron—1 : 3 : 3 arrangement; (*b*) monocapped trigonal prism—1 : 4 : 2 arrangement; (*c*) pentagonal bipyramid—1 : 5 : 1 arrangement; (*d*) 1 : 2 : 2 : 2 arrangement.

same. This will also affect the relative stabilities of the structures considered and might account for the apparent preference for the 1 : 4 : 2 rather than the 1 : 2 : 2 : 2 structure. The bonds having the greatest number of close neighbouring bonds would be predicted to be longer than the others, as in the case of the axial bonds in the trigonal bipyramid. In the above three cases the bonds to the ligands located on the central circle would be expected to be longer than the others. Also, in the case of non-equivalent ligands, the more electronegative ligands which have the smaller bond-pairs would be expected to occupy these central positions and larger bond-pairs would occupy the remaining positions. Thus, for $UO_2F_5^{3-}$, which has a pentagonal bipyramid structure, the fluorine atoms are in the pentagonal plane as they are more electronegative than oxygen and because the uranium oxygen bond almost certainly has an order of two or higher. In NbF_7^{2-} we would predict that the bonds to the four equivalent fluorines would be longer than the bonds to the other fluorines. We also expect the bonds to the fluorines in the pentagonal plane of IF_7 to be somewhat longer than the bonds to the axial fluorines.

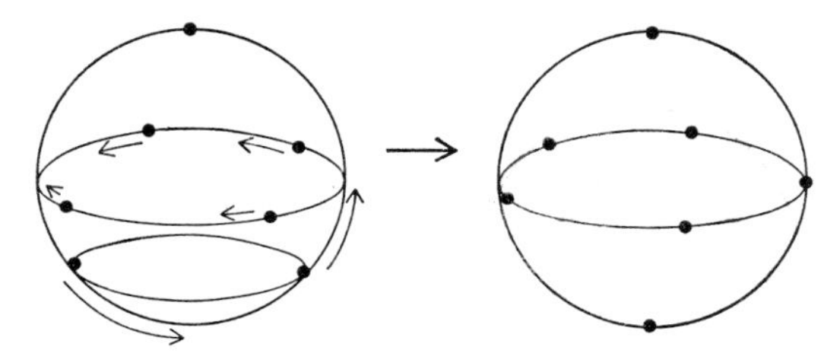

FIG. 5.4 Ligand motions required to convert the monocapped trigonal prism to the pentagonal bipyramid.

There is no information on the structure of any molecule of this type in solution, probably because of rapid intramolecular fluorine exchange similar to that proposed for trigonal bipyramid molecules. This type of intramolecular rearrangement can always occur easily when there are two or more structures with very similar energies, and when there is a low activation energy for the conversion of one to another. Figure 5.4 shows the motions of the ligands required to convert the monocapped trigonal prism into the pentagonal bipyramid. The ^{19}F n.m.r. spectra of IF_7 and ReF_7 in solution indicate complete equivalence of the fluorine atoms. As this is not possible except in a most unlikely planar molecule, we must conclude that there is rapid fluorine exchange probably occurring by an intramolecular mechanism.

The ions $TeCl_6^{2-}$, $TeBr_6^{2-}$, $SbBr_6^{3-}$, and other six co-ordinated halide complexes of Se (IV), Te (IV), and Sb (III), all have seven electron pairs in their valence shells, i.e., they are of the type AX_6E and they should have structures based on an arrangement of seven electron pairs. Although it is not possible to predict with complete certainty the most probable arrangement of seven electron pairs it is clear that the ligands would not be expected to have a regular octahedral arrangement, and yet all these ions do in fact have regular octahedral structures. It appears therefore that these molecules provide some of the few examples of a stereochemically inactive lone-pair. This is further discussed in Chapter 7.

Although TeF_6^{2-} is not known, the related species IF_6^- and XeF_6 appear to have the expected non-octahedral structures based on a valence shell of seven electron pairs, including a lone-pair. There has been some discussion as to the extent that XeF_6 deviates from an octahedral structure, and it could be that the distortion is rather small because of a tendency for the non-bonding pair to be stereochemically inactive as in $TeBr_6^{2-}$.

5.3 EIGHT ELECTRON PAIRS

The preferred arrangement for eight electron pairs that maximizes their distance apart is the square antiprism. Although the cube is a more symmetrical polyhedron, electron-pair repulsions are clearly greater in the cube than in the square antiprism. The square antiprism may be derived from the cube by rotating one square face by 45° with respect to the opposite face, which increases all the distances between the corners of opposite faces. As is general for high co-ordination numbers, other polyhedra must have rather similar energies. The rules given in Section 5.1 concerning the arrangement of mutually repelling points on the surface of a sphere lead to the arrangements shown in Fig. 5.5: (*a*) the triangular dodecahedron (bisdisphenoid) or 2:2:2:2 arrangement; (*b*) the bicapped trigonal prism or 2:2:4 arrangement; (*c*) the bicapped trigonal antiprism, puckered hexagonal bipyramid or 1:3:3:1 arrangement; and (*d*) the hexagonal bipyramid or 1:6:1 arrangement in addition to the square antiprism (*e*) and cube (*f*) as possible alternative polyhedra for eight co-ordination. As the cube is evidently less stable than the square antiprism it is not expected that it will be observed as a structure for a simple discrete AX_8 molecule in which the X are simple unidentate ligands.

Table 5.3 lists all the known structures of compounds of the type AX_8, i.e., discrete molecular species with unidentate ligands. Except in one case the observed structure is the square antiprism. The octacyanomolybdate (IV) ion has the dodecahedral (bisdisphenoid) structure in the solid state, but it has been claimed from infra-red

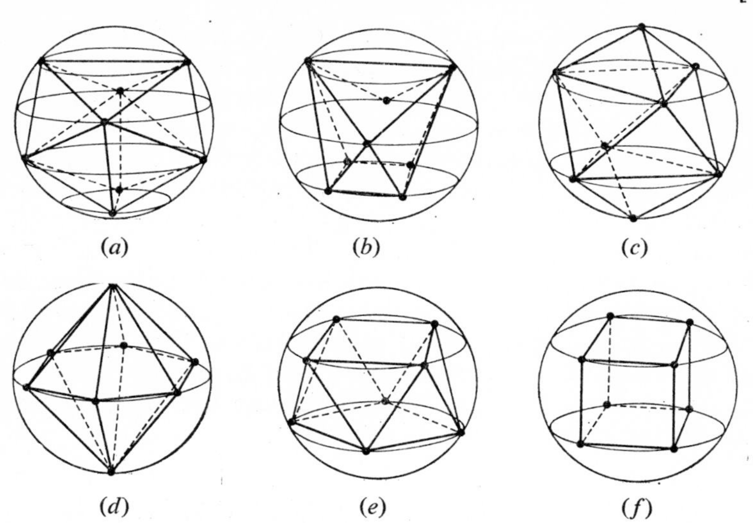

FIG. 5.5 Possible arrangements of eight electron pairs: (*a*) Triangular dodecahedron on bisdisphenoid—2 : 2 : 2 : 2 arrangement; (*b*) Bicapped trigonal prism—2 : 2 : 4 arrangement; (*c*) Bicapped trigonal antiprism—1 : 3 : 3 : 1 arrangement; (*d*) Pentagonal bipyramid—1 : 5 : 1 arrangement; (*e*) Square antiprism—4 : 4 arrangement; (*f*) Cube—4 : 4 arrangement.

and Raman spectral studies that in solution the $Mo(CN)_8^{4-}$ ion has the square antiprism structure. This does not seem to be well-established, but if it is the case it would appear that packing considerations favour the dodecahedron (bisdisphenoid) in the crystal. It should be

Table 5.3

Antiprism	Dodecahedron
TaF_8^{2-}	$Mo(CN)_8^{4-}$
ReF_8^{2-}	
$Sr(H_2O)_8^{2+}$	
$Ba(H_2O)_8^{2+}$	
$Eu(H_2O)_6Cl_2^{+}$	
$Gd(H_2O)_6Cl_2^{+}$	

noted however that molybdenum (IV) has a d^2 inner shell which is asymmetrical and could possibly therefore affect the stereochemistry, although in general such a small number of d electrons does not have any observable effect (see Chapter 8).

Uranyl complexes with bidentate ligands, e.g. $UO_2(NO_3)_3$, adopt a puckered hexagonal bipyramid structure which may be regarded as being dictated by the UO_2 group which has two large multiple bonds which will take up more space in the valency shell than the single bonds to the nitrate groups and which have a strong tendency to adopt a collinear arrangement.

5.4 NINE ELECTRON PAIRS

The predicted tricapped trigonal prism is the only structure that has been observed for nine-co-ordination. The known examples of AX_9 molecules having this structure are ReH_9^{2-}, TcH_9^{2-} $Nd(H_2O)_9^{3+}$, $Er(H_2O)_9^{3+}$, $Y(H_2O)_9^{3+}$, and $Sc(H_2O)_9^{3+}$.

No examples are known of molecules with simple unidentate ligands having more than nine electron pairs in the valence shell.

REFERENCES AND SUGGESTIONS FOR FURTHER READING

R. J. GILLESPIE, *Can. J. Chem.*, **38,** 818 (1960); **39,** 2336 (1961).

S. J. LIPPARD, *Progress in Inorganic Chemistry* Vol. 8 (Edited by F. A. Cotton), p. 109, Interscience (1967).

E. L. MUETTERTIES and C. M. WRIGHT, *Quart. Rev. Chem. Soc.*, **21,** 109 (1967).

6

Molecular Geometry of the Second Period Elements (Lithium to Neon)

6.1 CORE SIZE AND CO-ORDINATION NUMBER

In this chapter we consider the geometry of the molecules of the second period (Li to Ne) in terms of the principles outlined in the previous chapters. It is convenient to treat these elements before the remainder of the elements in the main groups of the periodic table for several reasons. The most important is that the valence shells of these elements are small and the number of electron pairs that can be accommodated in the valence shell is limited. As a consequence, the elements rarely exhibit co-ordination numbers greater than four, whereas higher co-ordination numbers are common for the remaining main group elements.

On moving from left to right in any period of the periodic table, the size of the core decreases as the nuclear charge increases, and hence the maximum number of electron pairs that can be accommodated in the valence shell decreases, as is illustrated by the data in Table 2.1. At the same time, the strength of attraction of the core for electron pairs increases with increasing charge, and it thus tends to surround itself with an increasing number of electron pairs. The number of electron pairs that a core will attract is approximately that number which is needed to balance the core charge: counting -2 for an unshared pair and -1 for a shared pair (or somewhat less than -1 for a pair shared with an electronegative ligand such as F. Thus the atom as a whole tends to carry only a small, or zero, charge in accordance with the electroneutrality principle first enunciated by Pauling. The metallic elements on the left of the periodic table have large cores and therefore large valence shells which can always accommodate more than the number of electron pairs that they tend to attract. Thus, in the third period (Na to Ar), the maximum co-ordination number of aluminium for electron pairs is eight, whereas it requires only three pairs to balance its core charge. Because this leaves the valence shell rather empty, aluminium may commonly acquire four pairs of electrons as in $AlCl_4^-$, or as many as six with the very electronegative ligand fluorine as in the ion AlF_6^{3-}. The valence shells of silicon and phosphorus can accommodate up to six electron pairs, and there is therefore ample room for the four or five shared pairs that they require to neutralize the core charge. With an electronegative ligand they acquire the maximum possible six electron pairs as in the ions SiF_6^{2-} and PCl_6^-. The smaller valence shell of sulphur is barely large enough to accommodate six electron pairs, and chlorine can only accommodate four—unless the ligands are rather electronegative. Hence we find SF_6 but not SCl_6, and the only compounds in which chlorine has more than four electron pairs in its valence shell are those with the very electronegative ligands, oxygen and fluorine. Finally, as the valence shell of argon can contain only four electron pairs—even if these are binding electronegative ligands such as fluorine—and as the valence shell is already filled by the four electron pairs in the neutral argon atom it shows no tendency to form any compounds. The same two opposing tendencies of decreasing core size limiting the number of electron pairs that can be accommodated in a valence shell and the tendency of the increasing core charge to increase the number of electron pairs attracted into the valence shell are, of course, operative in every period. For the second period elements (Li to Ne) these two opposing tendencies come sufficiently into conflict that they lead to the elements nitrogen, oxygen, and fluorine having some most unusual properties that are not encountered elsewhere in the periodic table.

Lithium, like all the metals, has a rather large core and therefore has a valence shell that can accommodate a rather large number of electron pairs. Since its small core charge of +1 enables it to attract rather few electron pairs its valence shell is never complete, and its co-ordination number for electron pairs is never limited by the size of the valence shell. In its compounds it may have as many as six, or, in principle, even more, surrounding ligands, but the bonding is predominantly ionic and each ligand donates only a small amount of charge to the central lithium.

Beryllium needs two shared electron pairs to neutralize its core charge, but as its valence shell may easily accommodate four, and probably as many as six, electron pairs, it tends to form complex ions such as $BeCl_4^{2-}$ and BeF_4^{2-} in which it acquires four electron pairs with rather electronegative ligands. Boron requires three electron pairs and is limited to four by the size of its valence shell, hence it forms molecules such as BF_3 and complexes such as BCl_4^-. With carbon however the size of the valence shell definitely limits the number of electron pairs in the valence shell to four, which is the number of shared pairs required by carbon to balance its nuclear charge. Thus carbon forms an almost unlimited number of compounds in which it has four covalent bonds, but not a single molecule of the type CX_5^- is known. The smaller valence shells of nitrogen and oxygen can however accommodate only three electron pairs—at least, according to the hard-sphere model—although their core charge requires five and six shared pairs in order to achieve neutrality. As these elements in their compounds all have four electron pairs in their valence shells it is evident that the intense electric field resulting from the high charge and small size of the core is sufficient to hold four electron pairs quite strongly, although there must evidently be some compression of the electron-pair orbitals or, alternatively, according to the hard-sphere model the electron pair spheres cannot be in contact with the central core. This unusual situation gives rise to a strong tendency for these elements to minimize the strain between these crowded electron pairs in any possible manner, and in turn leads to a number of unique features of the structural chemistry of these elements which are discussed in this chapter.

6.2 MULTIPLE BONDING

One of the unique features of the elements carbon, nitrogen, oxygen, and fluorine is that they have a much greater tendency to form double and triple bonds than any other elements. This can be attributed to their high electronegativities resulting from their highly charged cores and to the strong repulsions between the electron pairs in their valency shells. Double and triple bonds are only formed by relatively electronegative elements such as carbon, nitrogen, oxygen, and fluorine that can attract two or more electron pairs strongly enough to hold them in the bonding region despite their mutual repulsions. Moreover as a double-bond orbital takes up somewhat less space than two single bond orbitals and a triple bond orbital somewhat less space than three single bond orbitals there is an additional tendency for these elements to minimize electron pair interactions in their crowded valency shells by forming multiple bonds whenever possible. For example carbonic acid is $O{=}C(OH)_2$ whereas silicic acid is $Si(OH)_4$, and carbonyl compounds are not known to polymerize although the analagous silicon compounds, the silicones, are singly bonded polymers.

$$\begin{array}{l} R \\ \quad \diagdown \\ \quad\quad C{=}O \\ \quad \diagup \\ R \end{array} \qquad\qquad \begin{array}{ccccccc} & R & & R & & R & \\ & | & & | & & | & \\ - & Si & -O- & Si & -O- & Si & -O \\ & | & & | & & | & \\ & R & & R & & R & \end{array}$$

Whereas unsaturated hydrocarbons containing carbon–carbon double bonds are numerous and stable, no examples are known of silicon–silicon multiple bonds in the analogous silanes. Silicon does not have a crowded valence shell when it contains only four electron pairs and therefore shows no tendency to double bond formation which, since it causes two electron pairs to be somewhat squashed together, would lead to a less stable structure than the corresponding structure containing two single bonds. Whereas it would appear that for carbon, nitrogen, and oxygen the small loss in energy due to the crowding together of electron pairs in the double bond is more than compensated by the reduction in the interelectronic repulsions in the crowded valence shell.

The crowding of electron pairs in the valence shells of nitrogen, oxygen, and fluorine shows itself dramatically in the bond energies of the N—N, O—O, and F—F bonds which have long been known to be anomalously low (Table 6.1). The bond energies would be expected to increase somewhat in the series C—C $<$ N—N $<$ O—O $<$ F—F

Table 6.1 Bond energies (kcal)

—C—C—	—N̈—N̈—	:Ö—Ö:	:F̤̈—F̤̈:
83	38	33	37
>C=C<	—N̈=N̈—	:Ö=Ö:	
148	100	96	
—C≡C—	:N≡N:		
194	226		
C—H	N—H	O—H	F—H
99	93	111	135
Si—Si	P—P	S—S	Cl—Cl
42	41	63	58

because of the increasing charge of the core, and the corresponding increased attraction for the bond electron-pair, as they do in the next period from Si—Si to Cl—Cl, and in the series C—H, N—H, O—H, F—H. In fact they decrease rather dramatically, and this can be ascribed to strong electron-pair repulsions between the crowded valence shells of the two atoms. These are particularly important when the valence shell contains one or more large lone-pairs as it does for nitrogen, oxygen, and fluorine. The interactions between the lone-pairs on the adjacent atoms are primarily responsible for the weakness of these bonds. Thus the bond strength decreases markedly from C—C to N—N which has adjacent lone-pairs, and on passing from the N—N bond to the O—O bond there is a further decrease in strength despite the increasing core charge. This can be attributed to the presence of the two non-bonding pairs on each oxygen which offset the expected increase in strength. Finally, in the fluorine molecule, there is a slight increase in strength as the increasing core charge has a small effect in increasing the bond energy despite the increase in the number of lone-pairs.

In the C—H, N—H, O—H, and F—H bonds, as hydrogen has no lone-pairs, there is no interaction between lone-pairs on adjacent atoms, and the bond energy increases in the expected manner with increasing nuclear charge. For the C—C, C=C, and C≡C bonds the bond energy of a double bond is less than twice that of a single bond, and the bond energy of the triple bond less than three times that of a single bond. This may be attributed to the increased repulsion between the electron pairs when two or three pairs are crowded into the bonding region. For N—N bonds however this effect is offset by the diminishing repulsion between lone-pairs on the adjacent nitrogen atoms as they move apart in the series $>\ddot{N}-\ddot{N}<$ $-\ddot{N}=\ddot{N}-$, :N≡N: and accordingly the bond strength increases by approximately three on passing from the single to the double bond and again by a factor of approximately two on passing to the triple bond. Because of the lone-pair : lone-pair repulsion the N=N bond is weaker than the C=C bond but the N≡N bond is stronger than the C≡C bond because the lone-pairs are now on the opposite sides of the molecule and too far apart to interact strongly with each other.

The attraction of the cores of carbon, nitrogen, oxygen, and fluorine for electrons is so strong, and the repulsion between four pairs in their valence shells so strong, that even triple-bond formation occurs readily for these elements, particularly when bonded to each other. Thus acetylene, hydrogen cyanide, nitrogen, and carbon monoxide all contain triple bonds. The strong attraction of the two small, highly charged cores for electron pairs is able to crowd three pairs into the bonding region, thus relieving the repulsions between the electron pairs in each of the valence shells to some extent.

$$\mathrm{H{-}C{\equiv}C{-}H} \quad :\mathrm{N{\equiv}N}: \quad \mathrm{H{-}C{\equiv}N}: \quad :\mathrm{N{\equiv}\overset{+}{O}}: \quad :\mathrm{C{\equiv}O}:$$

6.3 STABLE MOLECULES CONTAINING UNPAIRED ELECTRONS

Another unusual feature of the elements carbon, nitrogen, oxygen, and fluorine is the occurrence of stable molecules of these elements containing unpaired electrons, i.e., stable-free radical molecules such as NO and NO_2 which both have one unpaired electron and O_2 which has two unpaired electrons. The occurrence of these unusual

$$:\dot{\mathrm{N}}{=}\ddot{\mathrm{O}}\dot{\cdot} \qquad\qquad :\mathrm{N}{=}\dot{\mathrm{O}}:$$

(1) (2)

Table 6.2 Triple bond lengths

—C≡C—	1·20 Å
—C≡N:	1·16
:N≡N:	1·10
:C≡O:	1·13
:N≡O:+	1·06
˙{:O≡O:}+	1·12
˙{:O≡O:}˙	1·21
˙{:N≡O:}	1·15

molecules may again be attributed to the strong tendency of these elements to form multiple bonds. The bond length of the NO molecule is only 1·15 Å which is comparable to that of a number of triple bonds (Table 6.2) and is certainly considerably less than the value of 1·22 Å normally given for an N=O double bond (Table 1.7). However, if the structure of NO is written in a conventional manner so as to allow no more than eight electrons in the valence shell of either atom then the molecule contains a double bond and either nitrogen or oxygen can have only seven electrons in its valence shell (1) and (2). It seems more reasonable therefore to write a structure for NO with a triple bond. However, if this is done one electron is left over and there remains the problem of accommodating this electron in the molecule. It may be seen in Figure 6.1 that the concentration of much of the electron density of the molecule into the triple bond leaves a region with rather little electron density outside the bonding region which it seems reasonable to assume has some residual affinity for electrons. Since this lies largely outside the normal valence shell it may be described as a secondary valence shell, and

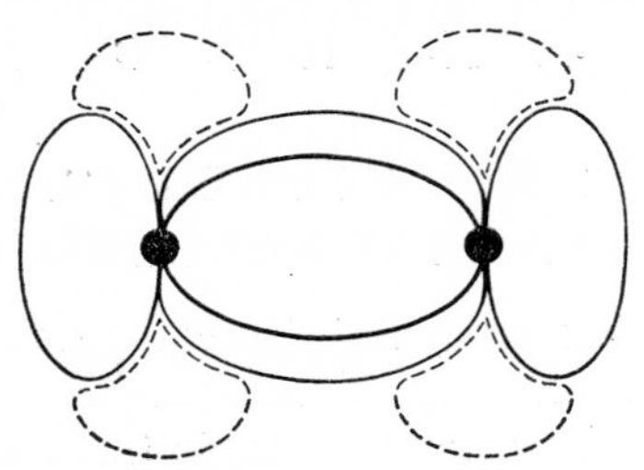

FIG. 6.1 The secondary valence shell ------- in molecules containing a triple bond, e.g., N_2, CO, NO, and O_2.

since it lies outside the bonding region it is somewhat anti-bonding in character. It appears to be a special feature of molecules of the elements C, O, N, and F that they can accommodate one or two electrons in this secondary valence shell.

$$^{\cdot}\{:N{\equiv}O:\} \qquad ^{\cdot}\{:O{\equiv}O:\}^{+} \qquad ^{\cdot}\{:O{\equiv}O:\}^{\cdot}$$

(3) (4) (5)

In writing formulae for such molecules we place the electrons in the ordinary valence shell inside curly brackets and any electrons in the outer secondary valence shell outside these brackets as in (3), (4) and (5) which show formulae for NO, O_2^+ and O_2 written in this manner. If it is assumed that this secondary valency shell is able to accommodate two electrons with parallel spins then the para-magnetism of the oxygen molecule is accounted for. The existence of this secondary valency shell for the CO and N_2 molecules is also demonstrated by the co-ordination complexes formed by carbon monoxide and by nitrogen with transition metals (6) which appear to owe their stability to their ability to accept non-bonding electrons into this secondary valency shell (see p. 199).

The fact that oxygen has the triple bond structure (5) rather than the double bond structure (7) indicates that in this case the repulsions between the electron pairs in the oxygen valence shell are minimized

$$\ddot{M}{-}C{\equiv}O: \qquad :\!\ddot{O}{=}\ddot{O}\!:$$

(6) (7)

by allowing three electron pairs to come under the attraction of two nuclei and at the same time promoting two electrons to the secondary valence shell.

6.4 MOLECULAR GEOMETRY OF THE ELEMENTS LITHIUM TO FLUORINE

The possible electron-pair arrangements and the resulting stereochemistries for the elements lithium to fluorine based on a valence shell containing four electron pairs are summarized in Table 6.3. Four electron pairs give rise to tetrahedral, pyramidal, and angular molecules for the cases of zero, one and two non-bonding pairs respectively. As lithium, beryllium, and boron have only one, two, and three electrons in their valence shells they frequently have in their compounds fewer than four electron pairs in their valence shells. In such cases three bonding pairs give a planar trigonal molecule, and two bonding pairs a linear molecule. The possible stereochemistries for molecules containing multiple bonds are also included in this table. Not all formally possible cases are included, e.g., cases where the central atom acquires an improbably high formal charge have generally been omitted. In those cases where the central atom has a formal charge of ± 2, the bonding will be expected to have a considerable amount of ionic character so that the actual charge on the atom is considerably smaller.

6.5 LITHIUM

Lithium has a large central core and therefore can accommodate a large number of electron pairs in its valence shell (nine according to the hard-sphere model), but as its low charge of $+1$ enables it to attract only a few electron pairs its valence shell is always incompletely filled. The vast majority of lithium compounds are crystalline solids in which the bonding is regarded as predominantly ionic. Moreover, since lithium is obviously unable to have lone-pairs in its valence shell the same stereochemistry would arise from the electrostatic repulsion of ligands in ionic bonding or from the mutual repulsion of bonding electron pairs in covalent bonding. Many lithium salts are known in hydrated forms in which the lithium ion is surrounded by the expected tetrahedral arrangement of four water molecules; occasionally, as in $LiClO_4.3H_2O$, there are six water molecules around each lithium in an octahedral arrangement (8). In $LiOH.H_2O$, the structure of which is shown in Fig. 6.2, there are four

Table 6.3 Molecular geometries for the elements beryllium to fluorine

Single bonds

Number of electron pairs in the valence shell	2	3	4		
Number of lone-pairs	0	0	0	1	2
Shapes	linear	trigonal planar	tetrahedral	pyramidal	angular
	—Be—	Be^-	Be^{2-}		
	$—B^+—$	B	B^-		
		C^+	C	C^-	
			N^+	N	N^-
			O^{2+}	O^+	O
					F^+

Multiple bonds

Number of 'orbitals'	2	3	
Number of lone-pairs	0	0	1
Shapes	linear	trigonal planar	angular
	=B—	B^-	
	—C≡	C	
	=C=		C^-
	$—N^+≡$		
	$=N^+=$	N^+	N
	$—O^{++}≡$		
	$=O^{++}=$	O^{++}	O^+

(8)

oxygen atoms in a tetrahedral arrangement around each lithium, each water molecule bridging between two lithiums. In lithium hydride there is an octahedral arrangement of hydrogens around

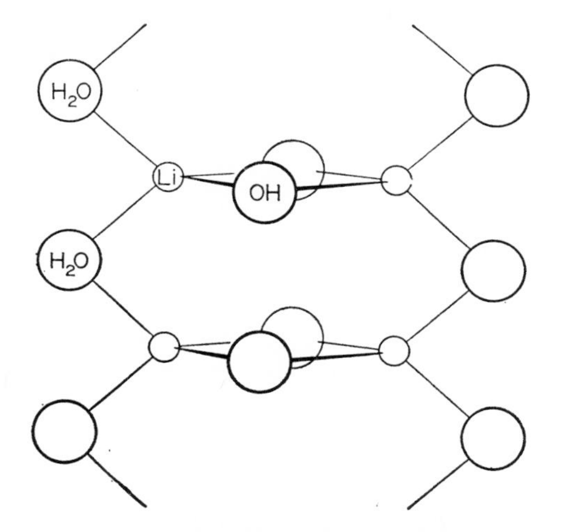

FIG. 6.2 The structure of $LiOH.H_2O$.

each lithium. There is probably some covalent character in the bonding, and in any case, since a hydride ion is essentially an electron pair orbital containing an embedded proton, there are in fact six electron pairs with an octahedral arrangement occupying the valence shell of the lithium.

6.6 BERYLLIUM

The beryllium halides BeF_2, $BeCl_2$, $BeBr_2$, and BeI_2 have been found to have the expected linear structure in the vapour state. The vapours of beryllium chloride and bromide have been found to contain dimers also. These presumably have a square structure with three co-ordinated beryllium (9).

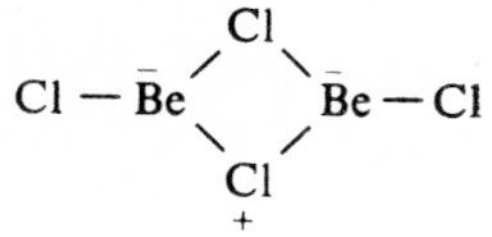

(9)

As the valence shell of beryllium is incompletely filled in its ordinary divalent BeX_2 compounds it has a strong tendency to acquire electron pairs in the formation of complexes. The data in Table 2.1 suggests that beryllium just has room for six electron pairs in its valence shell. However, the acquisition of four additional electron pairs to form a complex BeX_6^{4-} would place a very large charge on

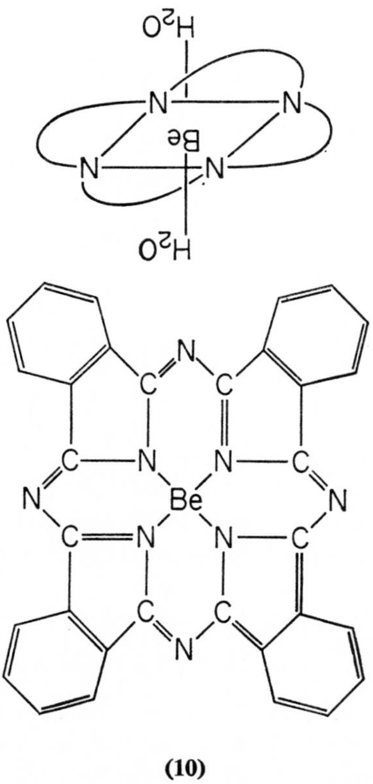

(10)

beryllium, and no examples of this type of complex are known except the apparently unique beryllium phthalocyanine dihydrate, in which beryllium has the expected octahedral co-ordination by four coplanar nitrogens of the phthalocyanine molecule and two water molecules (10). As the water molecules form very polar bonds, and as the nitrogens may be held sufficiently far away from the beryllium by the rigid structure of the phthalocyanine molecule that they do not transfer an excessive amount of charge to the beryllium, it seems possible that in this way the total charge transferred to the beryllium is not more than it can comfortably accommodate, and thus the

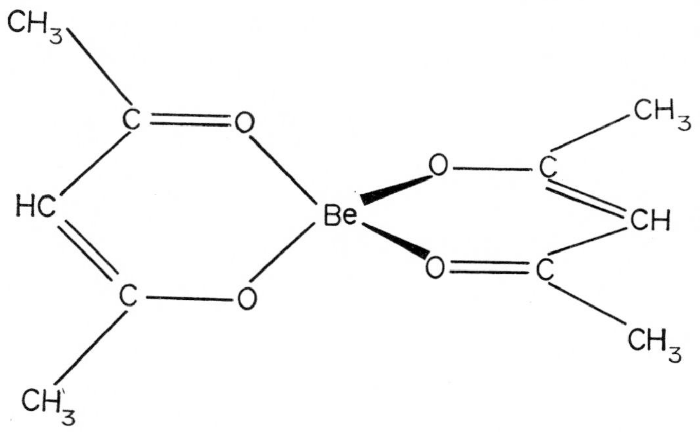

(11)

complex is stable. It would appear to be worthwhile to search for more six-co-ordinated beryllium compounds. For example beryllium hydride is evidently a polymeric material which might well contain six-co-ordinated beryllium.

However, in the vast majority of its compounds it is clear that beryllium is able to hold only four electron pairs in its valence shell and it forms many tetraco-ordinated complexes which in all cases have the expected tetrahedral structure, e.g., BeF_4^{2-}, beryllium acetylacetone (11), beryllium dichloride etherate $BeCl_2.2Et_2O$ (12) and the hydrated beryllium ion $Be(H_2O)_4^{++}$.

In the solid state beryllium dichloride has a continuous chain structure in which each beryllium is surrounded by four chlorine atoms giving approximately square four-membered $BeCl_2Be$ rings (13). The bond angles cannot be tetrahedral in such a four-membered

(12)

ring, and the experimental values are ClBeCl = 98·2° and BeClBe = 81·8°. It is unlikely that the angle between electron pairs in the beryllium valence shell could be as small as 98·2°, and it is

(13)

reasonable to assume that the directions of the electron pairs around beryllium do not coincide exactly with the Be—Cl directions, i.e., it is assumed that the bonds are somewhat bent. This would appear to

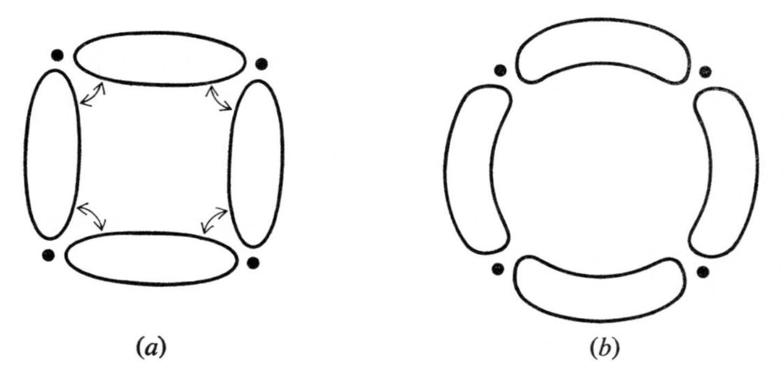

FIG. 6.3 Bent bonds in the square $BeCl_2Be$ group and similar systems (*a*) repulsions between bonds in a square $BeCl_2Be$ group and similar systems; (*b*) bond-bond repulsions minimized by bond bending.

be a rather common feature of small ring systems and can be attributed to the repulsions between the bonding electron pairs which force them away from the bond directions (Fig. 6.3). If in beryllium chloride each bond is bent only 5° from the Be—C direction then the angle between the electron pairs on beryllium is tetrahedral. The same amount of bond bending at chlorine would give a bond angle of 92°, and although the bending could perhaps be greater, the angle between electron pairs in the valence shell of chlorine may be as small as 92° because considerable distortion from the ideal angle of 109·5°

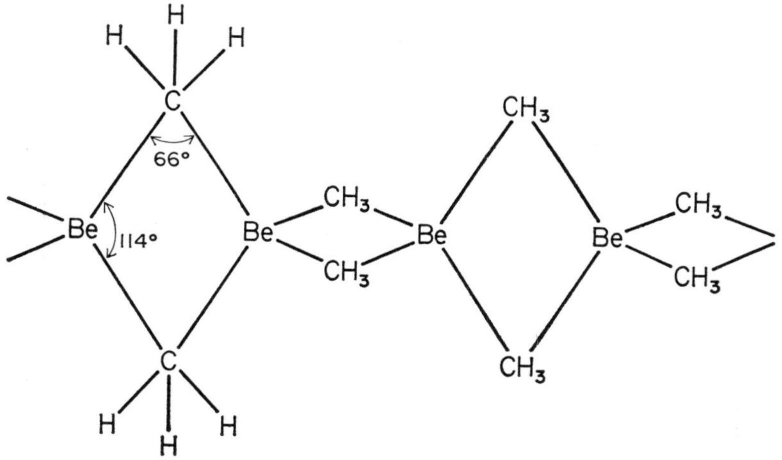

(14)

is possible by virtue of the relatively large size of the chlorine valence shell.

Dimethylberyllium has a very similar polymeric chain structure, although we note that the bond angles are rather different, i.e., CBeC = 114° and BeCBe = 66° (14). These differences reflect the different bonding in this compound which is electron deficient as there are not enough electrons to form ordinary two-electron bonds between each adjacent pair of atoms. Each carbon is bonded to two berylliums by one electron pair in a three-centre orbital. In this case the maximum electron density lies inside the Be—C direction and again an approximately tetrahedral arrangement of four electron pairs around both beryllium and carbon is maintained (Fig. 6.4).

Beryllium oxide has the wurtzite structure in which each beryllium is surrounded tetrahedrally by four oxygen atoms (Fig. 6.5), and although this compound might be regarded as ionic there must be

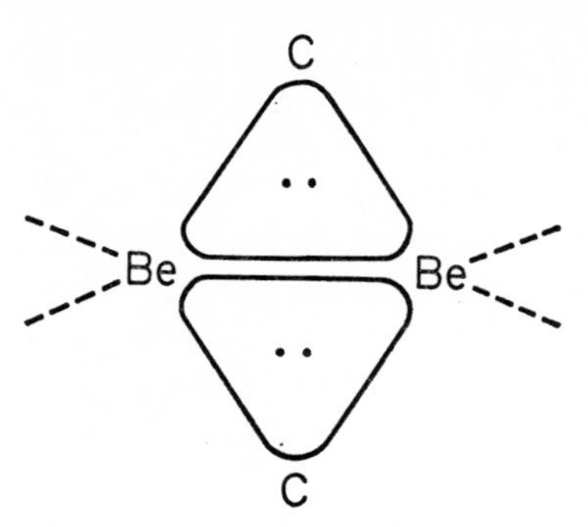

FIG. 6.4 Three-centre orbitals in beryllium dimethyl. Each beryllium is surrounded by an approximately tetrahedral arrangement of three such orbitals.

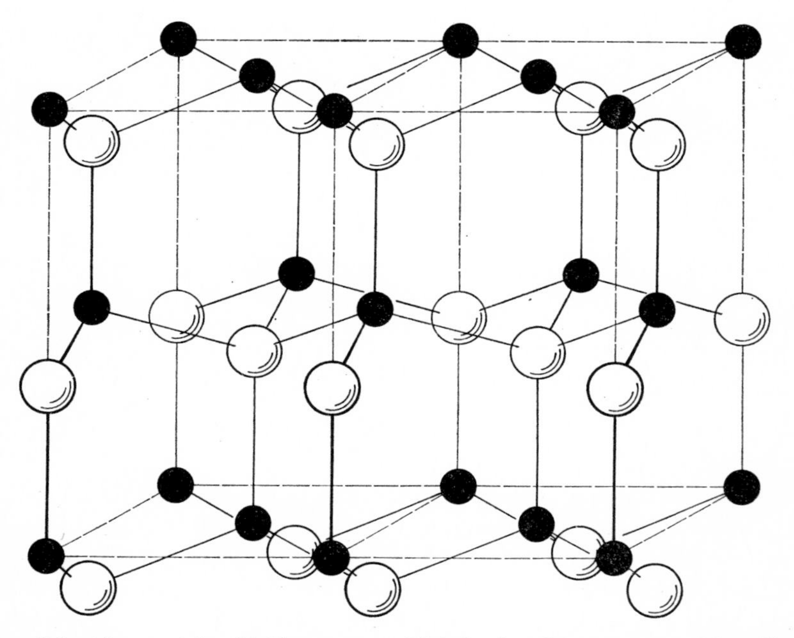

FIG. 6.5 The wurtzite (ZnS) structure. This is also the structure of beryllium oxide.

considerable sharing of electrons. Similarly, in beryllium fluoride, which has the β-cristobalite structure (Fig. 6.6), each beryllium is bonded tetrahedrally to four fluorines. In beryllium oxide nitrate and beryllium oxide acetate each beryllium is tetrahedrally co-ordinated in interesting structures in which a central oxygen atom is

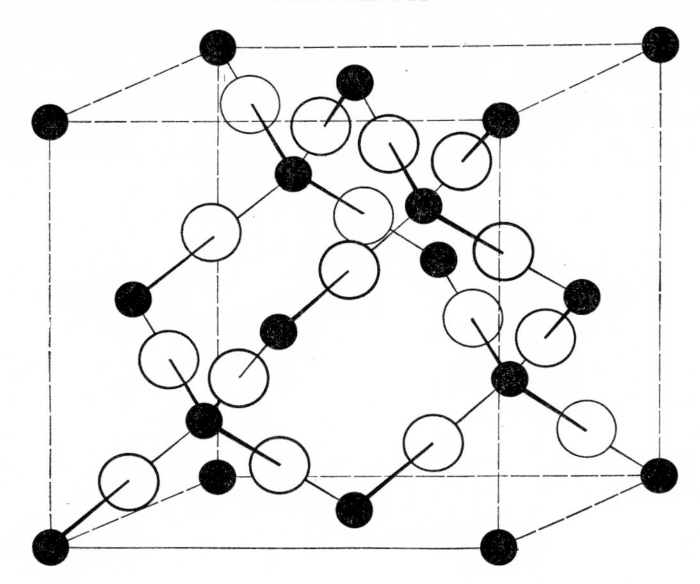

FIG. 6.6 The β-cristobalite (SiO_2) structure. This is also the structure of beryllium fluoride.

surrounded tetrahedrally by four berylliums, and the edges of the tetrahedron thus formed are bridged by nitrate or acetate groups (15). For simplicity, only three of the six chelating acetate groups are shown. The three remaining groups are opposite the three bottom edges of the tetrahedron.

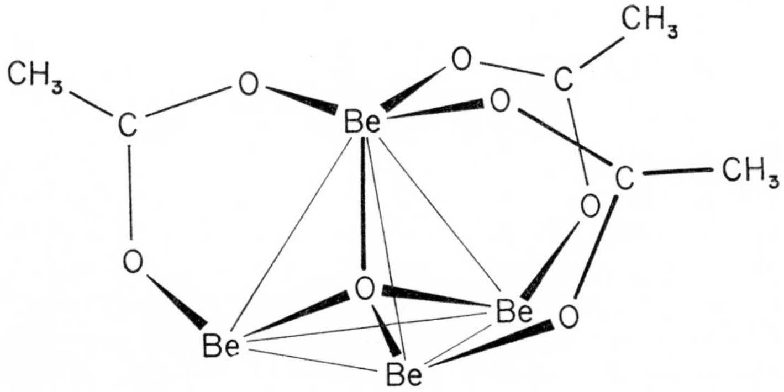

(15

6.7 BORON

The neutral boron atom forms three trigonal planar bonds, and a number of simple BX_3 compounds ($X = F, Cl, Br, CH_3, OCH_3$) have been shown to have a planar triangular shape in the gas phase.

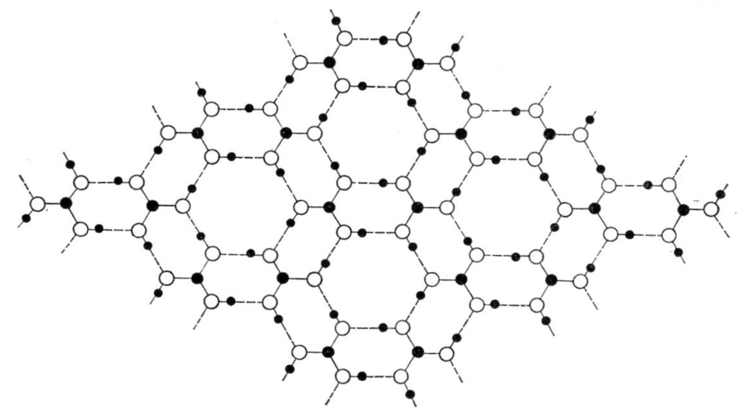

FIG. 6.7 Structure of H_3BO_3. Portion of one of the hydrogen bonded layers. ------ hydrogen bonds.

Orthoboric acid $B(OH)_3$ also has a trigonal planar arrangement of OH groups about the boron in the crystal (Fig. 6.7), the molecules being held together in sheets by hydrogen bonds. In both CH_3BF_2 and $C_6H_5BCl_2$ the bond angles are not all equal and the smallest

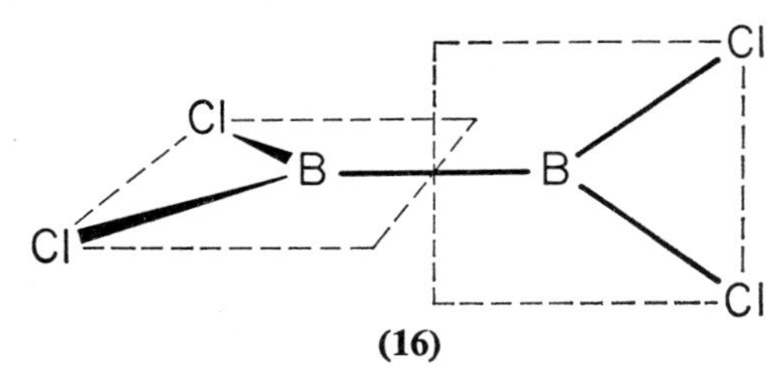

(16)

angle, 118° in both cases, is between the most electronegative groups, i.e., between the two fluorines or the two chlorines.

In the diboron tetrahalides, boron again has a trigonal planar arrangement of its bonds. An interesting feature of these molecules is that the BCl_2 groups are coplanar in the crystal but have a staggered configuration in the vapour (16). Repulsion between the B—X bonding pairs of electrons would be expected to lead to the staggered conformation and it would seem that crystal packing considerations lead to the planar form in the solid state.

A number of cyclic molecules are known in which boron has a trigonal arrangement of its three bonds, e.g., methyl boronic acid anhydride (17) and borazole (18). Boron nitride has an infinite planar structure (19) containing the same B—N ring as in borazole. In these molecules the B—N bond lengths indicate that there is

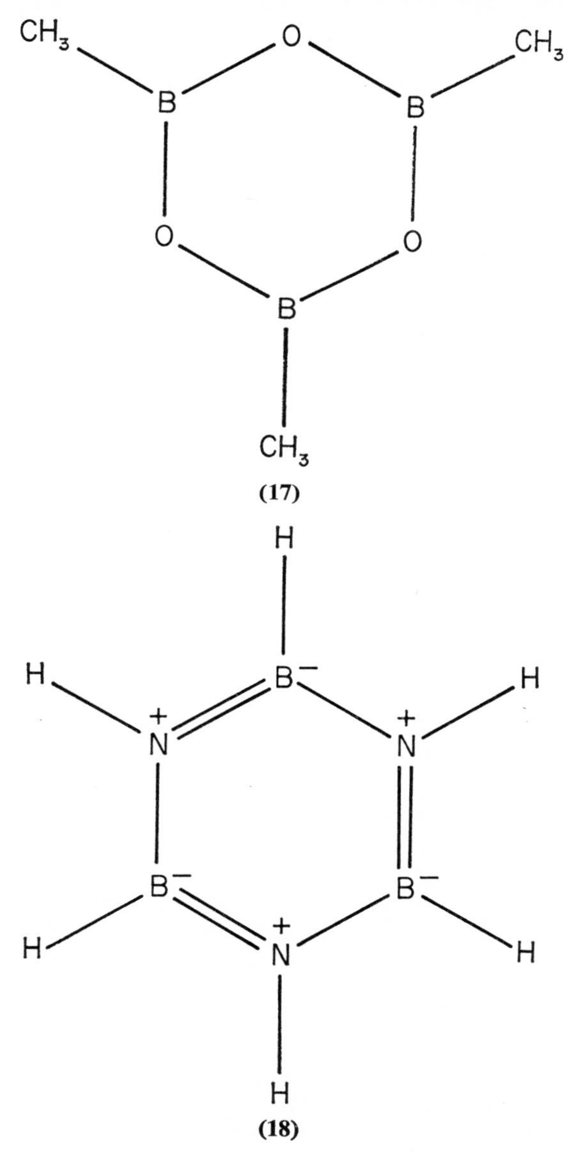

(17)

(18)

double bonding resulting from donation of nitrogen lone-pair electrons to the electron deficient boron. Similar double bonding may also be present in the cyclic anhydride (17). These double-bond electrons are best regarded as occupying molecular orbitals extending over the whole ring system, and they have no direct effect on the molecular geometry which is determined by the planar trigonal

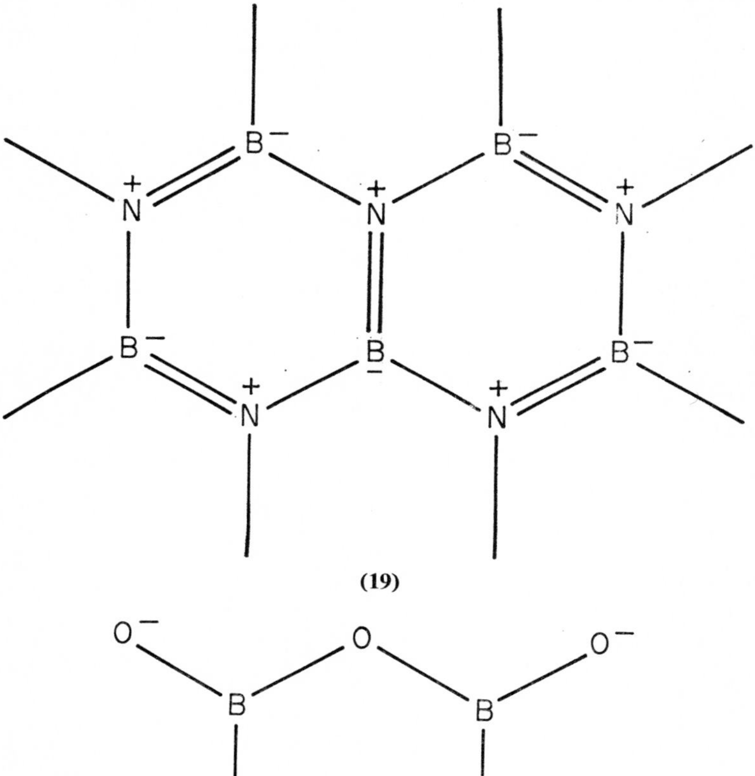

(19)

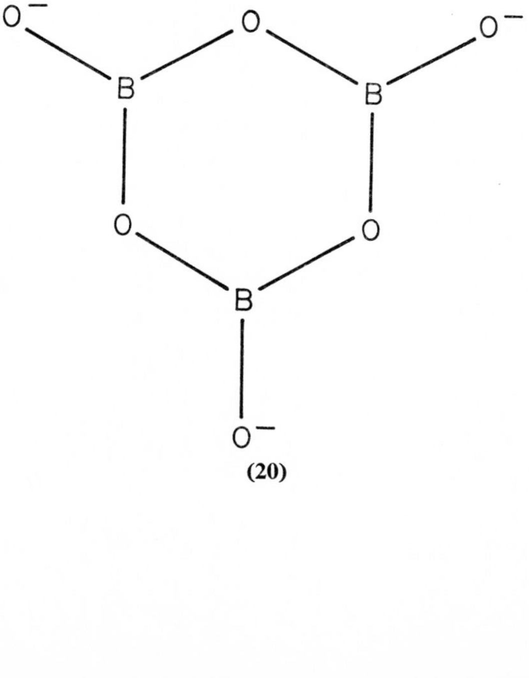

(20)

arrangement of three localized electron pairs in the valence shell of boron as is discussed below for graphite. Similarly, the metaborate ion (20) and metaboric acid have planar cyclic structures. The polymeric metaborate ion $(BO_2)_n^{n-}$ has an infinite linear chain structure in CaB_2O_4 (21) and a large BOB bond angle of 130° which must

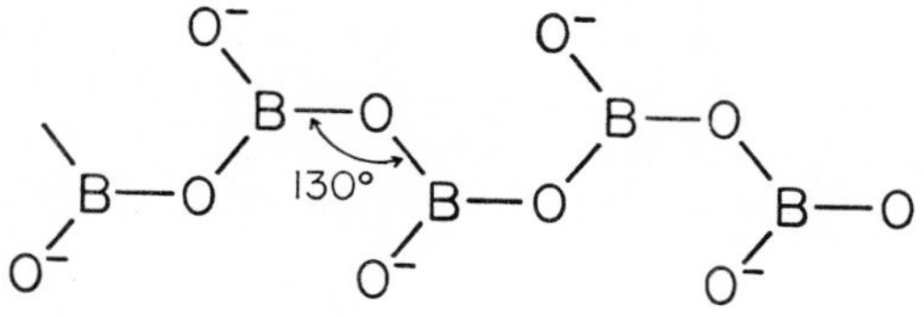

(21)

arise from delocalization of non-bonding electrons on oxygen into the vacant orbital on boron, i.e., the structure would be better represented as in (22).

The very strong tendency of boron to complete the octet in its valence shell is, of course, evident in the many tetrahedral complexes that it forms, e.g., BH_4^-, BF_4^-, $B(OH)_4^-$, $F_3B.NH_3$, and $F_3B.O(CH_3)_2$.

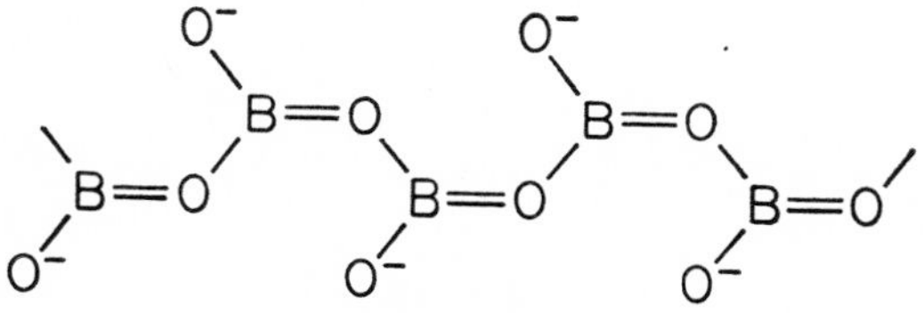

(22)

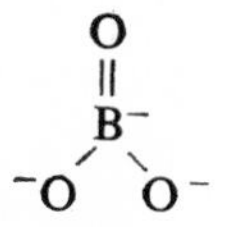

The possibility of both planar trigonal co-ordination and tetrahedral four co-ordination leads to the structures of borates often being quite complex. Only a very few compounds are known that contain the simple orthoborate anion BO_3^{3-}. In contrast to the carbonate ion the borate ion has a strong tendency to polymerize. In this property it resembles the silicate ion. Because of the smaller electronegativity of boron the double bonded structure does not contribute as much to the actual structure as the corresponding structure for the carbonate ion

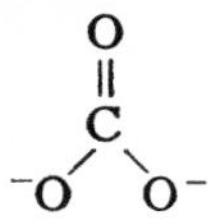

Thus the borate ion is unstable by virtue of the high charge on each oxygen and the incomplete valence shell of the boron. It is interesting to note that the only stable orthoborates have small polarizing cations Co^{2+} and Mg^{2+} which are able to remove some of the excess charge

(23)

(24)

on the oxygens. Typical examples of borates containing both three and four co-ordinated boron are $K[B_5O_6(OH)_4].2H_2O$ (23) and borax $Na_2[B_4O_5(OH)_4]8H_2O$ (24).

The structure of B_2H_6 has been discussed in Chapter 3 (Fig. 3.20). The observed geometry is consistent with three-centre bridge bonding, the smaller bond angle between the bridging bonds than between the terminal BH bonds being due to the smaller electron density in

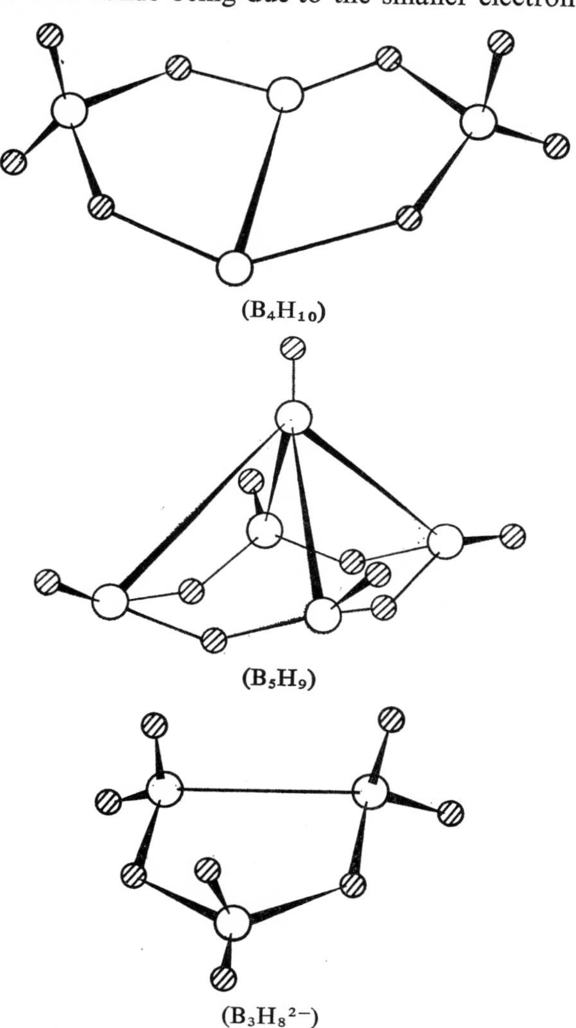

FIG. 6.8 Structures of the boron hydrides B_4H_{10} and B_5H_9 and the anion $B_3H_8^{2-}$.

the BH bridging bonds. The B_4H_{10} molecule and the $B_3H_8^{2-}$ anion (Fig. 6.8) can similarly be described in terms of two-centre bonds and three-centre B—H—B bridge bonds. The interesting tetrahedral molecule B_4Cl_4 has been discussed in Chapter 3. It also provides an example of three-centre bonding.

The higher boron hydrides and their anions involve still more delocalized bonding and all the electrons cannot be described in terms of localized pairs. For example, in B_5H_9 all the terminal and bridging hydrogens can be bonded by localized two-centre and three-centre bonds respectively, leaving six electrons for the boron–boron bonding in this square pyramid. No satisfactory description of the boron–boron bonding in this square pyramid can be given in terms of localized electron pairs.

6.8 CARBON

(a) Tetrahedral CX_4 Molecules

Bond angles at carbon in a number of simple CX_4 compounds are listed in Table 6.4. The most important point about these data is that all the deviations from the tetrahedral angle are very small. This is a consequence of the fact that four bonding pairs of electrons are

Table 6.4 Bond angles at a tetrahedral carbon atom

	HCH	XCH	XCX		HCH	XCH	XCX
CH_3F	110·3°	108·3°		$CHBr_3$		108·0°	110·8
CH_2F_2	111·9		108·3°	CHI_3			113·0
CHF_3		110·5	108·8			FCF	FCCl°
CH_3Cl	110·5	108·5		$CClF_3$		108·6	110·2
CH_2Cl_2	112·0	104·7	111·8	$CFCl_3$		107·3	111·5
$CHCl_3$		108·6	110·4			ClCCl	
CH_2ClF	111·9	109·1	110·0	CCl_2F_2		113	110
CH_3Br	111·2	107·6		CH_3OH	109·3°		
CH_3Cl	110·5			CH_3NH_2	109·5		
CH_4I	111·4	107·4		CH_3SH	109·8		

closely packed around the small carbon atom, and changes from the tetrahedral angle are resisted by strong electron-pair repulsions. When the ligands are hydrogen and fluorine the angle between the fluorines is always the smallest angle which is consistent with the greater electronegativity of fluorine and the consequently smaller size of the C—F bonding electron pair and this is also the case when the ligands are fluorine and chlorine. However, when the ligands are chlorine and hydrogen the chlorine bond angle is the largest, despite the greater electronegativity of chlorine. These unexpected bond angles can presumably be attributed to the effects of ligand–ligand repulsions. With a small central atom such as carbon, and a relatively large ligand such as chlorine, repulsions between lone-pairs on adjacent chlorine atoms are expected to become of some importance. Moreover, as has been mentioned previously (p. 58), bond angles involving hydrogen are generally smaller than would be predicted on the basis of its electronegativity alone.

(b) Pyramidal CX_3E Molecules

Carbanions CX_3^- are expected to have this pyramidal structure. However, the only structures that have been determined are those of rather stable carbanions with strongly electron withdrawing ligands such as NO_2 and CN and these are planar, e.g., $C(NO_2)_3^-$, or almost planar, e.g., $C(CN)_3^-$ due to extensive delocalization of the carbon lone-pair into the ligand orbitals. This may be described in terms of resonance structures such as

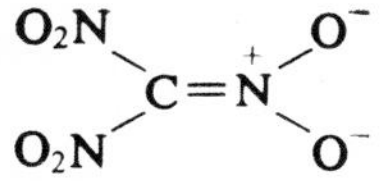

$$(O_2N)_2C{=}\overset{+}{N}(O^-)_2 \quad \text{and} \quad (:N{\equiv}C)_2C{=}C{=}\ddot{N}:^-$$

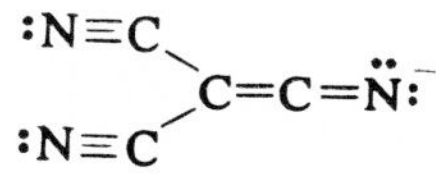

(c) Trigonal planar CX_3 Molecules

Carbonium ions R_3C^+ are expected to have planar trigonal structures although there is no direct evidence for this, e.g., an X-ray determination of the structure of a crystal. Some rather indirect evidence for the postulated planar structure has been obtained from n.m.r. spectroscopy. Other CX_3 compounds must contain one double bond, e.g., $X_2C{=}O$, $X_2C{=}CY_2$.

Since three orbitals adopt a planar trigonal arrangement, and since the double-bond orbital has a slightly greater diameter in the molecular plane than the single-bond orbitals, the angles between the double bond and the single bonds are expected to be somewhat greater than 120°, while the angles between the two single bonds therefore will be slightly less than 120°. This is generally found to be the case and some examples are given in Table 6.5.

The carbonate ion which has three equivalent bonds and a symmetrical triangular structure with 120° bond angles has been discussed in Chapter 3. It may be represented as in (25) with three equivalent bonds. It is usually assumed that in accordance with the octet rule these bonds have a bond order of $1\frac{1}{3}$ so that the total bond order around carbon is four and each oxygen therefore carries a $-\frac{2}{3}$

Table 6.5 Bond angles in doubly-bonded carbon compounds

	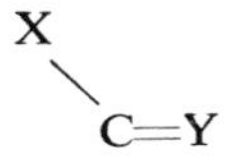 X–C=Y	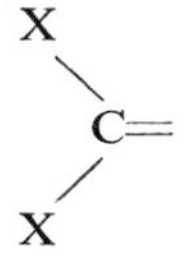 X–C–X (in $X_2C=$)	
$(CH_3)_2C=C(CH_3)_2$	124·3°	111·3°	
$(CH_3)_2C=CH_2$	122·3	115·3	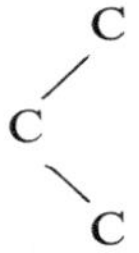 C–C–C
$Cl_2C=O$	124·3	111·3	
$F_2C=CH_2$	125·2	109·4	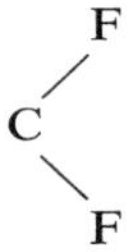 F–C–F
$CH_3HC=O$	123·9		
$F_2C=O$	126·0	108·0	
HFC—O		109·9	
$CH_3ClC=O$		112·7	
$H_2C=CH_2$		116·8	

charge. However, because each double bond in the plane of the molecule is only slightly larger than a single bond, it should be possible to pack the three double electron pairs of the three double bonds around carbon, particularly if they are somewhat polarized towards

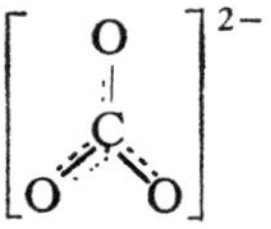

$$\left[CO_3 \right]^{2-}$$

(25)

the oxygens, and there is no reason to assume that when such multiple bonds are formed carbon necessarily strictly obeys the octet rule. It is interesting to note that the length of the C—O bond in the carbonate ion has been reported to have values varying from 1·24 to 1·29 Å.

From a bond-order bond-length plot a bond order of 1·6 to 1·9 can be deduced which is considerably greater than the bond order of 1·33 predicted by the octet rule.

Graphite consists of parallel planar layers of carbon atoms that are only weakly bonded together. In each layer each carbon atom is bonded to three others by two single and one double bond and the planar trigonal arrangement of these three bonds determines the

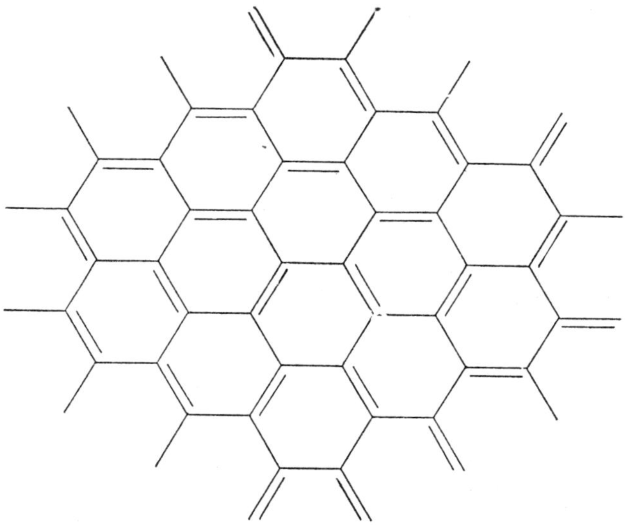

FIG. 6.9 One of the many possible resonance structures for part of one of the layers of carbon atoms in graphite.

planarity of each carbon atom layer. In fact all three bonds are equivalent and all the bond angles are equal to 120° as each double bond can occupy each of the three possible positions on each carbon atom and this gives rise to an enormous number of different resonance structures for a given layer (Fig. 6.9). This is equivalent to one pair of electrons from each double bond occupying delocalized orbitals covering the whole of each layer. This, however, has no effect on the geometry which is determined by the planar arrangement of the three localized single bonds on each carbon atom.

(d) Linear CX_2 Molecules

Molecules in which carbon forms two double bonds, e.g., carbon dioxide or ketene, $H_2C=C=CH_2$ or one triple and one single bond, e.g., HCN and acetylene are linear. In such linear molecules there is no reason to assume that the valence-shell electrons are closely paired, and presumably electrostatic repulsion between the two spin sets on each atom keeps them apart. Carbon dioxide may then be represented as in Fig. 6.10 and acetylene as in Fig. 6.11.

The C═O bonds in CO_2 are about 28 kcal more stable than would

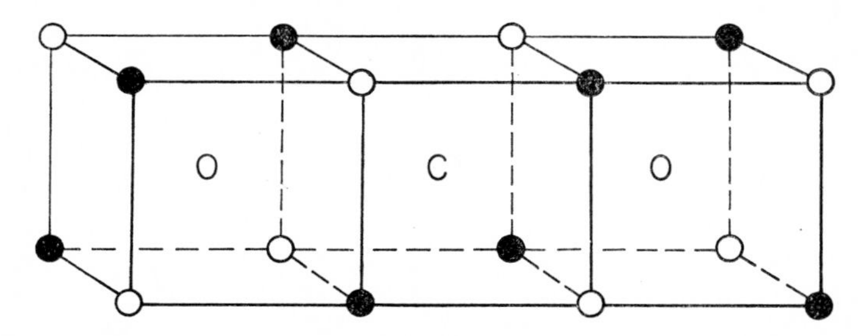

FIG. 6.10 Most probable relative configuration of electrons in CO_2. The two spin sets designated by ○ and ● respectively. Each set adopts a tetrahedral arrangement around each nucleus but the two sets are non-coincident.

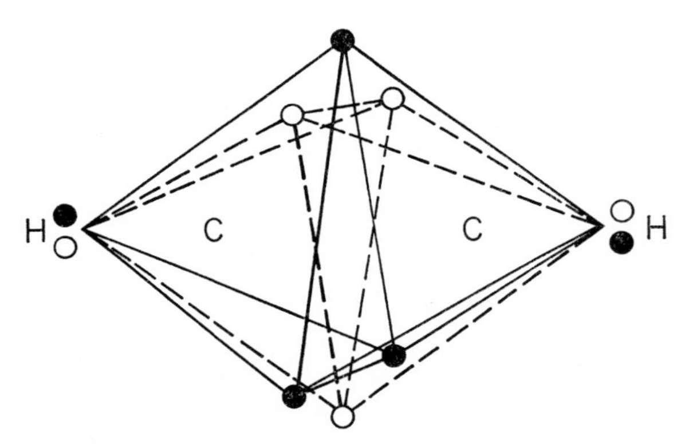

FIG. 6.11 Most probable relative configuration of electrons in acetylene. The two spin sets are designated by ○ and ●. There is a tetrahedral arrangement of each spin set around each carbon atom but the two sets are not coincident around the molecular axis.

be expected for normal C═O bonds as in H_2CO, they are stronger as measured by the force constant, and the bond length is only 1·16 Å which is considerably shorter than the normal C═O length of 1·23 Å as found in H_2CO and is almost as short as in CO which has a triple bond (Table 6·6). It seems reasonable to suppose that when the electrons in the valence shell of carbon are confined in two

Table 6.6 Carbon–oxygen bonds

	Length (Å)	$k \times 10^5$ (dyne cm^{-1})	Bond energy (kcal $mole^{-1}$)
C—O in CH_3OH	1·427	5·8	84
C=O in H_2CO	1·225	12·3	164
O=C=O	1·163	15·5	192
C≡O	1·131	18·6	257

double bonds some empty space is left in the valence shell of the central carbon in the region between the two double bonds, i.e., the secondary valence-shell is stabilized and concentrated mainly on the carbon atom. This secondary valence-shell can, to some extent, accept lone-pair electrons from oxygen which, because of the strong repulsion between the electron pairs in the crowded oxygen valence shell, have a strong tendency to delocalize into any adjacent available space, so that each bond acquires a certain amount of triple-bond

:O≡C≡O:

(26)

character (26). Another factor contributing to the stability and short bond lengths in CO_2 is no doubt the reduction in interelectronic repulsion that results from the fact that in a linear molecule it is not necessary for the electrons in the bonds to remain in close pairs, but the spin sets may separate as discussed above (Fig. 6.10).

6.9 NITROGEN

(a) Tetrahedral NX_4 Molecules

The expected tetrahedral arrangement of four single bonds about a positively charged nitrogen atom has been established for the ammonium ion and a number of tetraalkyl ammonium ions NR_4^+ and the NF_4^+ ion. The interesting molecule F_3NO has the expected tetrahedral structure.

(b) Pyramidal NX_3E Molecules

Bond angles in some typical molecules are listed in Table 6.7. These angles are consistent with the principles outlined in Chapter 3. A more electronegative ligand reduces the size of the bonding pair so that it exerts smaller repulsions, and hence the angles involving this ligand are smaller than the angles of bonds to less electronegative

Table 6.7 Bond angles in NX_3E moleclues

NH_3	107.3°
NF_3	102·1
NHF_2	102·9 (FNF)
NH_2CH_3	105·9 (HNH), 112·1 (CNH)
NH_2OH	107 (HNH)

ligands. The very small dipole moment of NF_3 of 0·2 D compared with the moment of 1·5 D of ammonia can be accounted for by attributing a rather large dipole moment to the lone-pair which opposes the large N—F bond dipole, giving a small resultant dipole.

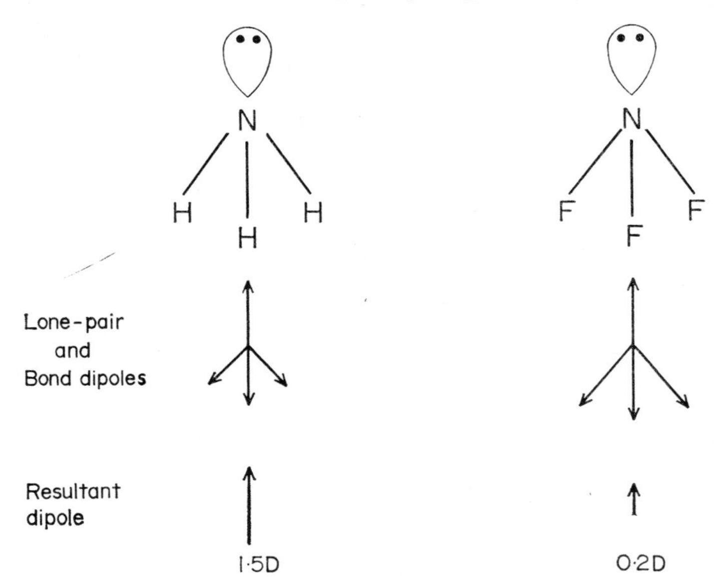

FIG. 6.12 Dipole moments of NH_3 and NF_3.

The NH dipole in ammonia is much smaller and does not compensate for the large lone-pair dipole; hence the molecule has a rather large resultant dipole moment (Fig. 6.12).

The molecules $N(SiH_3)_3$ and $N(GeH_3)_3$, however, have planar structures. This can be attributed to the very strong repulsions between the crowded electron pairs on nitrogen causing delocalization of the unshared electron pair onto silicon or germanium which have incomplete valence shells with only four electron pairs although they can accommodate at least six. Thus the structure of $N(SiH_3)_3$ can be written as in (27), the partial bonds indicating the delocalization

SiH_3 SiH_3
N
SiH_3

(27)

of the nitrogen lone-pair into the valence shell of the three surrounding silicons. The mutual interaction of the three N—Si bonds and the effective absence of a lone-pair on nitrogen causes the molecule to have a planar structure. This is another example of the special properties that arise from the crowding of electron pairs in the valence shell of the elements carbon to fluorine. In fact whenever these elements have lone-pairs in their valence shells and are bonded to heavier atoms with incomplete valence shells some delocalization of the lone-pairs into the vacant valence shell of the heavy atom

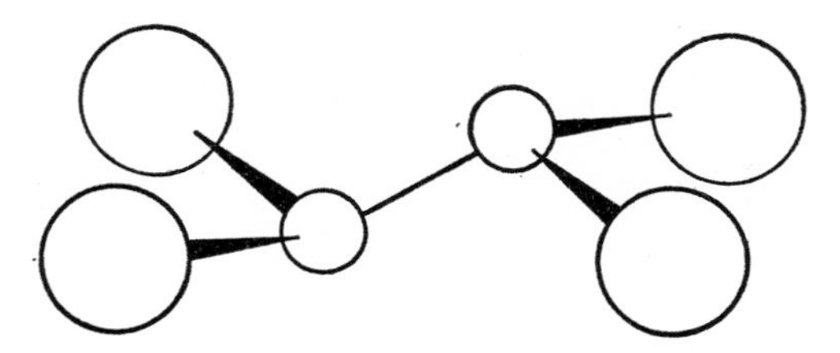

FIG. 6.13 The structure of N_2H_4 and N_2F_4.

occurs. Thus, in the related $O(SiH_3)_2$ molecule, the bond angle at oxygen is 155°, which is much larger than the approximately tetrahedral angle expected for a simple OX_2 molecule. However, in the molecules $S(SiH_3)_2$ and $S(GeH_3)_2$ in which the central sulphur atom has an incomplete rather than a crowded valence shell and therefore no tendency to delocalize its electrons the bond angles are respectively 100° and 98·9°.

In contrast to N_2F_4 and N_2H_4, which have the expected pyramidal arrangement of bonds at each nitrogen (Fig. 6.13), $N_2(CF_3)_4$ has a structure analogous to B_2Cl_4 (18) with an almost planar configuration around each nitrogen. It would appear that the explanation for this must again be that the nitrogen lone-pair is largely delocalized into vacant orbitals on the ligands. Although at first sight this would seem to be an unattractive explanation, as carbon also has a filled valence shell of four electron pairs, the data in Table 3.2 shows that the valence shell is probably not completely filled by the four bonding

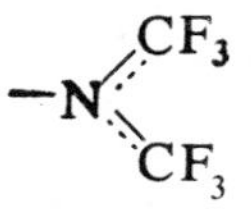

$$-N\begin{matrix} \cdot\cdot CF_3 \\ \cdot\cdot CF_3 \end{matrix}$$

(28)

electron pairs when three of them are to fluorine, and some delocalization of the nitrogen lone-pair onto the carbon may be possible. The bonding around nitrogen may then be represented as in (28) or alternatively in terms of structures of the type

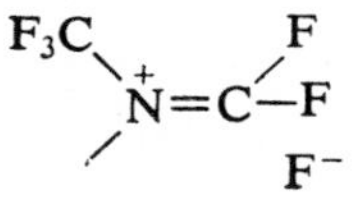

$$F_3C\diagdown\overset{+}{N}{=}C(F)_2\quad F^-$$

Of course fluorine also has a crowded valence shell with three lone-pairs which might also be expected to delocalize onto the carbon atom. This may indeed occur to some extent, but it appears that because the nitrogen lone-pair is held less strongly than the fluorine lone-pairs because of the lower electronegativity of nitrogen it delocalizes preferentially.

Similar effects are observed when the ligand, although having a

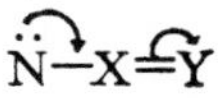

$$\ddot{N}{-}X{=}Y$$

(29)

filled valence shell, is part of an unsaturated group, i.e., is forming one or more multiple bonds to other atoms (29). In such a case X behaves as if it had an incomplete shell, since one of the electron pairs of the X=Y double bond may be partially transferred to Y, leaving X electron deficient and thus facilitating the transfer of lone-pairs on N into the N—X bond. This accounts for the planar 120° arrangement of the bonds around nitrogen in urea and formamide

(30). Although NX_3 molecules are usually pyramidal they are nevertheless rather easily distorted to the planar form. For ammonia, the activation energy required to reach the planar form is only about 6 kcal, consequently the molecule rapidly inverts. One can imagine either that the hydrogens flap up and down, or that the nitrogen

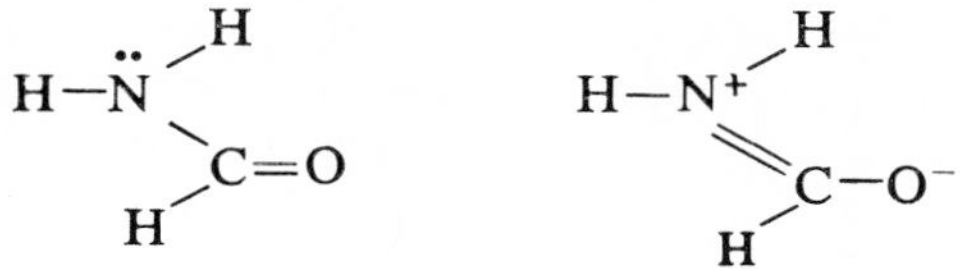

(30)

passes up and down through the plane of the hydrogens. The inversion frequency has been detected in the microwave spectrum. This rapid inversion of NX_3 molecules accounts for the fact that no optical isomers of NXYZ molecules have ever been isolated.

(c) Angular NX_2E_2 Molecules

The NH_2^- amide ion has the expected angular shape like the water molecule.

(d) Planar NX_3 Molecules

When nitrogen forms only three bonds and has no lone-pair, the three bonds, one of which must have more or less double-bond

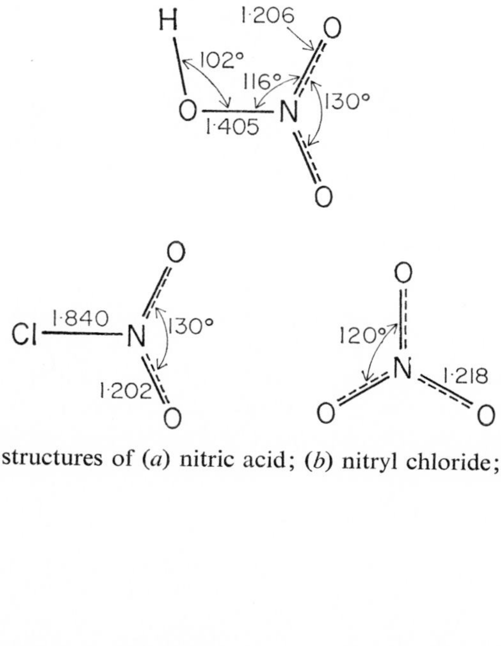

FIG. 6.14 The structures of (*a*) nitric acid; (*b*) nitryl chloride; (*c*) nitrate ion.

character, always adopt the expected trigonal planar arrangement, e.g., in the nitrate ion, nitric acid, nitryl chloride, and nitryl fluoride (Fig. 6.14).

All these molecules have surprisingly short NO bond lengths when compared with the normal single bond length of 1·37 Å and a double bond length of approximately 1·20 Å. It must be concluded that the bond order is approximately two in each case giving a total bond order of approximately five for N in HNO_3, NO_2F, and NO_2Cl. The N—F and N—Cl bonds are also surprisingly long compared with the normal single bond lengths of 1·34 and 1·69 Å respectively. The NO_2 portion of the molecule in both cases is remarkably similar to the NO_2 molecule (40). The bonding in these two molecules thus closely resembles that in the nitrosyl halides, the single unpaired electron in the secondary valency shell of the nitrogen in NO_2 being used to form a long bond with fluorine or chlorine, e.g.

O
Cl:{N
O

(31)

The nitrate ion is conventionally described by resonance structures such as

O
N+
O O−

which give an NO bond order of $1\frac{1}{3}$ and a total bond order of 4 at nitrogen and leave nitrogen with a positive charge. If however we make use of Pauling's electroneutrality principle according to which the atoms in a molecule tend to be neutral we can write the following as one possible resonance structure

O
N
O O−

the three equivalent structures then giving a NO bond order of $1\frac{2}{3}$ and a total bond order at nitrogen of 5. This higher bond order is much more consistent with the bond length of only 1·22 Å. As in the carbonate ion it is possible for the central carbon or nitrogen to accommodate this number of electron pairs in its valence shell provided some of the electron pairs have their volume decreased by being incorporated in multiple bonds. Nitrogen in fact shows a strong tendency to achieve a valence shell containing the five shared electron pairs that it needs to balance its core charge. A particularly interesting example is the tetrahedral molecule F_3NO in which the NO bond length of only 1·16 Å clearly indicates a bond order of not less than two again giving a total bond order of approximately five at nitrogen. Some further examples are discussed below.

(e) Angular NX_2E Molecules

Molecules of this type are expected to be angular, with a bond angle of approximately 120°, or probably slightly less than this because of the presence of the lone-pair. The structures of the molecules N_2F_2 and azomethane $CH_3N{=}NCH_3$ with bond angles of 115±5° and 110±10° respectively, are in agreement with this expectation. Both molecules exist as *cis* and *trans* isomers.

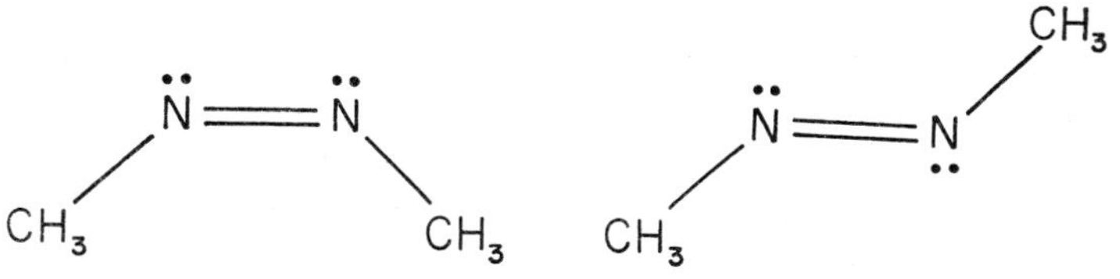

The nitrosyl halides and nitrous acid also have an angular structure with bond angles of somewhat less than 120° (Table 6.7). However, in the nitrosyl halides the N—halogen bonds are unexpectedly long and the N—O bond is shorter than expected, and in fact is slightly shorter than in NO where we have proposed that the bond is a triple bond with a single electron in the secondary valence shell. The ONX molecules may perhaps be best represented by a structure in which

Table 6.8 Bond lengths and bond angles in the nitrosyl halides

X	Angle	Bond lengths (Å) N—O	N—X (obs.)	N—X (calc.)*
F	110°	1·13	1·52	1·34
Cl	114°	1·14	1·95	1·69
Br	114°	1·15	2·14	1·84

* From single bond covalent radii.

the single electron in the secondary valence shell of NO forms a bond with the halogen. As this electron is at a greater average distance from the nucleus than the electrons in the valence shell it is reasonable that it forms a longer bond, and that the N—O bond length remains essentially unchanged.

(f) Linear NX_2 Molecules

Typical molecules of this type are the nitronium ion NO_2^+, the azide ion N_3^-, and nitrous oxide N_2O. Conventional octet rule structures for these molecules are shown in (32), (33), and (34)

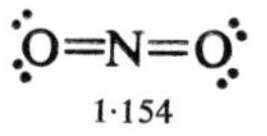

$$\ddot{O}{=}\overset{+}{N}{=}\ddot{O} \quad (1{\cdot}154) \quad \textbf{(32)}$$

$$\overset{-}{\ddot{N}}{=}\overset{+}{N}{=}\overset{-}{\ddot{N}} \quad (1{\cdot}15) \quad \textbf{(33)}$$

$$\overset{-}{\ddot{N}}{=}\overset{+}{N}{=}\ddot{O} \quad (1{\cdot}126 \;\; 1{\cdot}186) \quad \textbf{(34)}$$

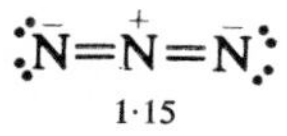

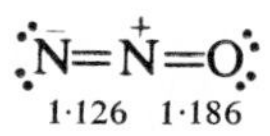

respectively, together with the observed bond lengths. These structures indicate charges in the molecules that appear to be unrealistically large, e.g., the N_2O molecule has a dipole moment of only 0·17 D, and they do not account for the fact that the observed bond lengths (except for the NO bond length in N_2O) are shorter than would be expected for double bonds. Using the electroneutrality principle and reducing the charge on each atom as far as possible, we can write structures (35), (36), and (37) in which the bonds in NO_2^+ and N_3^- have bond orders of 2·5 and the ionic charge becomes

$$\overset{+1/2}{:O}\equiv N \equiv \overset{+1/2}{O:} \quad \textbf{(35)}$$

$$\overset{-1/2}{:N}\equiv N \equiv \overset{-1/2}{N:} \quad \textbf{(36)}$$

$$:N{\equiv}N{=}O: \quad \textbf{(37)}$$

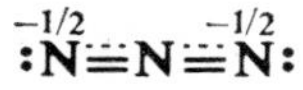

equally dispersed at each end of the molecule. Structure (37) containing a double and a triple bond is much more consistent with the dipole moment and the observed bond lengths than is (34). These structures again illustrate that more than four pairs of electrons can be accommodated in the valence shell of nitrogen provided that at least some of them are forming multiple bonds. Very similar molecules containing a linear N_3 group are hydrazoic acid (38) and methyl hydrazide (39) which bond lengths indicate should be written with a double and a triple bond as shown.

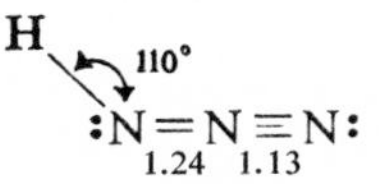

$$H\text{—}:N{=}N{\equiv}N: \quad (110^\circ;\; 1.24 \;\; 1.13) \quad \textbf{(38)}$$

$$CH\text{—}N{=}N{\equiv}N: \quad (125^\circ;\; 1.26 \;\; 1.13) \quad \textbf{(39)}$$

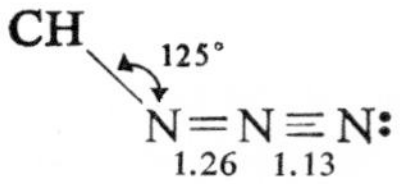

As all the electrons on nitrogen in NO_2^+ are involved in multiple bonding along the axis of the molecule it is reasonable to suppose that there is some vacant space in the vicinity of the nitrogen that can accommodate additional electron density, at least in a secondary valence shell. We may therefore write the structure of NO_2 as in (40) in which the NO bonds are essentially double bonds and are slightly longer than the bonds in NO_2^+. This structure is also consistent with the rather large bond angle of 134° and it gives a basis for understanding the structure of the rather unusual molecule N_2O_4 which is

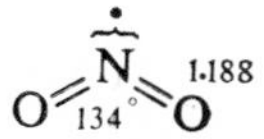

N 1.188
O 134° O

(40)

planar and has a very long N—N bond of 1·74 Å. The structure of N_2O_4 may be written as in (41). The N—N bond is long because it is formed by the unpaired electron on NO_2 which is the secondary valence shell and therefore at a greater average distance from the nitrogen nucleus than an electron in the primary valence shell. The formation of this bond has little effect on the remaining electrons in the molecule and the length of the NO bonds and the angle between them remain unchanged.

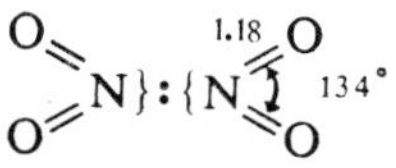

O 1.18 O
N}:{N 134°
O O

(41)

Because of the ellipsoidal shape of the NO double-bond orbitals and the rather large angle between the two double bonds the N—N-bond orbital will have a rather flattened disc-like shape and this serves to hold the molecule in a planar configuration (Fig. 6.15).

The bond angles at nitrogen in isocyanic acid, isothiocyanic acid,

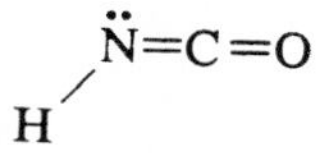

N=C=O
H

(42)

and the corresponding methyl compounds are surprisingly large. The lone-pair on nitrogen is clearly not exerting its full stereochemical effect in these molecules. It is presumably delocalized to some extent

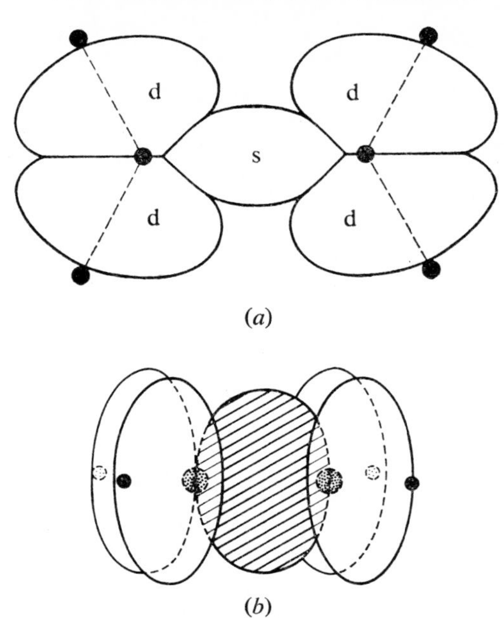

FIG. 6.15 Orbitals in the N_2O_4 molecule: (*a*) top view—d, double bond orbitals; s, the N—N single bond orbital formed by the electron in the secondary valence shell of each nitrogen; (*b*) side view, the N—N single bond orbital is shown shaded.

into the secondary valence shell of the carbon atom. The silyl H_3SiNCO and the derivatives ClSiNCO and Cl_3SiNCS are linear because of extensive delocalization of the nitrogen lone-pair into the partly empty valence shell of the silicon (43). The related H_3GeNCO molecule has a GeNC angle of 170°.

Table 6.9 Bond angles in $\ddot{N}=C=O$ molecules (X attached to N)

Molecule	Angle
HNCO	128·1°
CH_3NCO	125
HNCS	130·3
CH_3NCS	142

$$H_3\bar{Si}=\overset{+}{N}=C=O$$

(43)

6.10 OXYGEN

(a) Tetrahedral OX_4 Molecules

This is a rather unusual stereochemistry for oxygen in discrete molecules and in fact only very few examples are known, e.g., beryllium oxide acetate $Be_4O(CH_3COO)_6$ (15), zinc oxide acetate $Zn_4O(CH_3COO)_6$ and the nitrate $Be_4O(NO_3)_6$. As oxygen carries

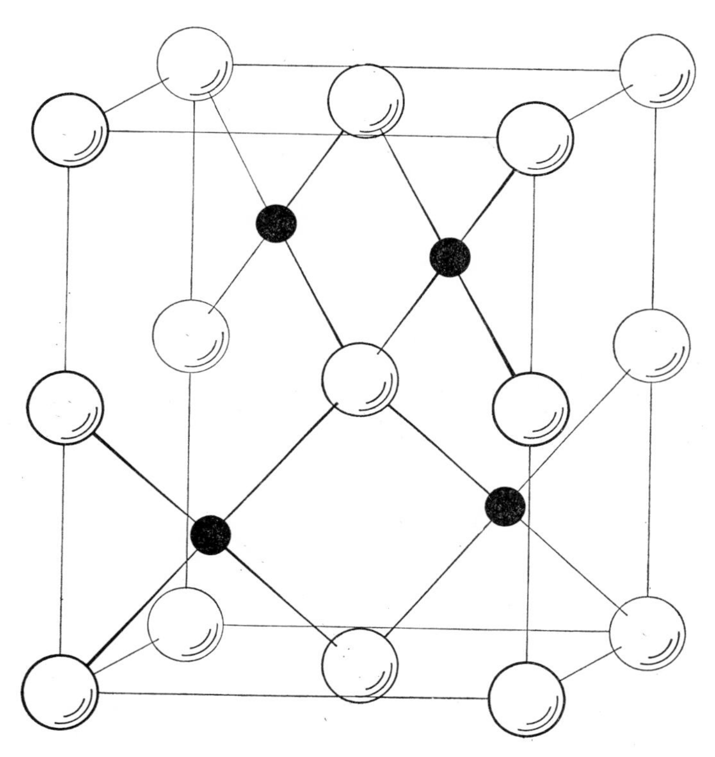

FIG. 6.16 The structure of PdO and PtO.

a formal double positive charge the bonds presumably have considerable ionic character. 50% ionic character would give zero charge to both the oxygen and the metal atoms. Many metal oxides have infinite lattice structures with the same tetrahedral arrangement of four bonds around oxygen. Although the bonds must have considerable ionic character there is nevertheless a tetrahedral arrangement of four electron pairs around oxygen, e.g., BeO and ZnO which have the structure shown in Fig. 6.5, PtO and PdO which have the structure in Fig. 6.16 and CuO and AgO which have a very similar structure. The structure of PdO differs considerably from those of ZnO and BeO because the metal atoms form four coplanar bonds rather than four tetrahedral bonds (Chapter 8). Because the metal and oxygen atoms thus form approximate squares in this structure (44) the PdOPd bond angle is only 98° but presumably there is some bond bending as in all small ring molecules.

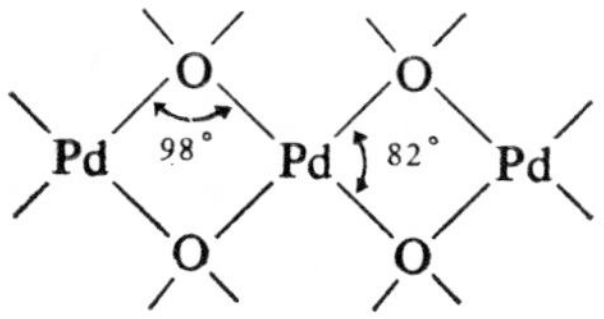

(44)

(b) Pyramidal OX_3E Molecules

The H_3O^+ ion and the adduct $F_3B—\overset{+}{O}(C_2H_5)_2$ are simple examples. However, when oxygen is attached to heavy atoms with incomplete

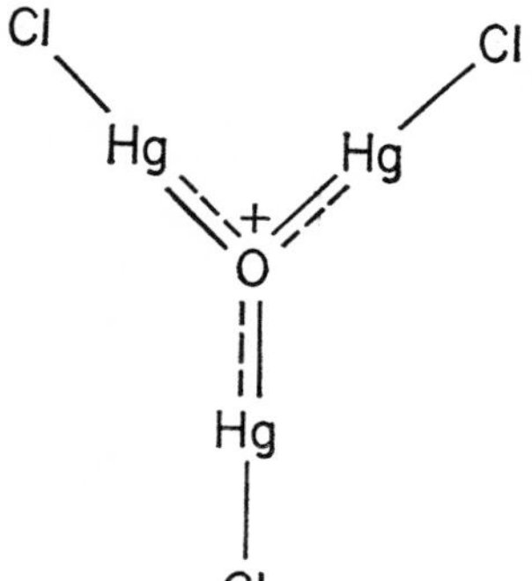

(45)

valence shells as in the ion $O(HgCl)_3^+$ (45) and the compound (46) the three bonds around oxygen are found to be coplanar because of delocalization of the oxygen lone-pair into the incomplete valence shell of the surrounding heavy atoms. This is quite analogous to the situation in the planar $N(SiH_3)_3$ molecule.

$$
\begin{array}{ccc}
(CH_3)_3Si & & \\
 & \diagdown & \\
 & O - Al(CH_3)_2 & \\
 & | \quad\quad | & \\
(CH_3)_2Al - O & & \\
 & \diagdown & \\
 & & Si(CH_3)_3
\end{array}
$$

(46)

(c) Angular OX_2E Molecules

The only clear example of a molecule of this kind is the ozone molecule (47). The bond angle of 117° is consistent with the presence of a lone-pair on the central oxygen and the short bond length indicates that the two bonds have considerable double-bond character

O (central, with lone pair); O–O 1.278; angle O–O–O 117°

(47)

(d) Angular OX_2E_2 Molecules

The molecules F_2O and H_2O have the expected small bond angles of less than 109·5°, the F_2O angle being less than the H_2O angle as expected from the greater electronegativity of fluorine (Table 6.10).

Table 6.10 Bond angles at oxygen in OX_2E_2 molecules

F_2O	103·2°	CH_2—O—CH_2 (ring, CH_2—CH_2 bonded)	61·6°
H_2O	104·5		
F_2O_2	109·5		
H_2O_2	96·9	CH_2—CH_2 / CH_2—O (four-membered ring)	94·5
$(CH_3)_2O$	111·5		

The bond angle in hydrogen peroxide is surprisingly small. It is possible that the hydrogen on one oxygen is attracted to an unshared pair on the other oxygen and that this tends to reduce the bond angle. The bond angle in dimethyl ether is slightly larger than expected. It is possible that this is due to a very slight delocalization of oxygen lone-pair electrons into the valence shells of the carbon atoms as discussed below for oxygen bonded to heavier elements.

The very small angles in ethylene oxide and trimethyleneoxide do not reflect the angles between the bonding pairs at oxygen, as the bonds must be rather bent. Whenever oxygen is attached to a heavy atom such as silicon or phosphorous considerable delocalization of the oxygen lone-pair electrons into the incomplete valence shell of the heavy atom occurs and the bond angle becomes greater than the tetrahedral angle. This may be seen from the data in Table 6.11, where bond angles at oxygen are compared with bond angles at sulphur in corresponding compounds. Without exception, the bond angles at oxygen are greater than tetrahedral, while the bond angles at sulphur are equal to or less than the tetrahedral angle. Sulphur, having an incomplete valence shell, does not have any tendency to delocalize its lone-pairs. It may be seen that as the size of the valence shell of the heavy atom decreases with decreasing size of the core from silicon to chlorine the bond angle at oxygen decreases as delocalization of the oxygen lone-pairs occurs to a decreasing extent. For silicon, which has considerable vacant space in its valence shell, the bond angle in many compounds reaches values as high as 150°. Any unsaturated system such as an aromatic ring can also accept electrons from oxygen, and again large bond angles at oxygen are observed. In some transition-metal complexes containing a bridging oxygen, delocalization of the oxygen lone-pairs may be essentially complete, giving rise to a linear oxygen bridge as in $[Cl_5RuORuCl_5]^{4-}$ and $[TiCl_2(C_5H_5)_2]_2O$.

It is unlikely of course that oxygen would carry a double positive charge, but presumably the bonds are polar with the electronegative oxygen at the negative end of the dipole, and this will reduce the positive charge on oxygen. A 50% ionic character in the bonds would give a neutral oxygen atom. These molecules can therefore be reasonably satisfactorily described by the two resonance structures

$$\overset{-}{X}=\overset{++}{O}=\overset{-}{X} \quad \text{and} \quad \overset{+}{X}\,\overset{--}{O}\,\overset{+}{X}$$

In the case of the linear Li_2O molecule the ionic structure is presumably the dominant one and, from electrostatic considerations alone, this would be expected to have a linear structure. There may, in addition, be a small contribution from the doubly bonded covalent structure—double bonds being formed because of the availability of vacant orbitals on the lithium. It should be noted that in a linear structure of this type there is no necessity for the two spin quartets on oxygen to be paired, and there may be a cubic arrangement of two tetrahedral sets of opposed spins on oxygen and lithium ions are then attracted to opposite sides of the cube (48).

$$\overset{+}{Li}\ \overset{--}{O}\ \overset{+}{Li} \qquad \overset{-}{Li}{=}\overset{++}{O}{=}\overset{-}{Li}$$

Table 6.11 XOX and XSX bond angles

SiOSi		SiSSi	
β-Quartz	142°	$(SiS_2)_n$	80°
β-Cristobalite	150	$((CH_3)_2SiS)_2$	75
$(SIO_3^{2-})_n$	137·5	$((CH_3)_2SiS)_3$	110
$(H_3Si)_2O$	144	$(SiH_3)_2S$	100
$((CH_3)_2SiO)_3$	136	$(SiH_3)_2Se$	96·6
$((CH_3)_2SiO)_4$	142		
POP		**PSP**	
P_4O_6	127·5°	P_4S_{10}	109·5°
P_4O_{10}	123·5	P_4S_3	103
$P_4O_{12}^{4-}$	132	P_4S_7	106
$(PO_3)_n^{n-}$	129	$P_4S_3I_2$	104
$P_2O_7^{4-}$	134		
$P_3O_{10}^{5-}$	121·5		
SOS		**SSS**	
$(SO_3)_3$	114°	S_8	105°
$(SO_3)_n$	121	S_n	100
$HS_2O_7^-$	114	S_4^{2-}	104·5
$S_3O_{10}^{2-}$	122	$(C_6H_4SO_2)_2S$	106·5
$S_2O_7^{2-}$	124	$(CH_3SO_2S)_2$	104
		$S_4O_6^{2-}$	103
ClOCl			
Cl_2O	111°		
Cl_2O_7	118·6		
GeOGe		**GeSSe**	
$(H_3Ge)_2O$	126·5	$(H_3Ge)_2S$	98·9
COC (aliphatic)		**COC (aromatic)**	
$(C_2H_5)_2O$	108°	$(p-IC_6H_4)_2O$	123°
$(CH_3)_2O$	111·5	$(p-BrC_6H_4)_2O$	123
1:4 dioxan	108	$(C_6H_5)_2O$	124
Paraldehyde	109·5	$p-C_6H_4(OCH_3)_2$	121

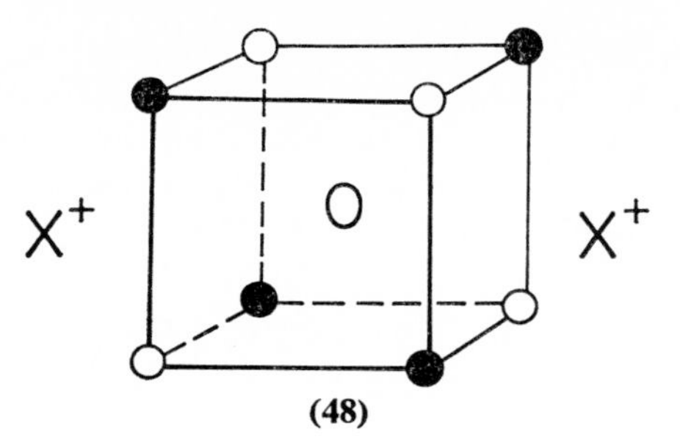

(48)

6.11 FLUORINE

Since fluorine very rarely forms more than one bond there is little to be said about its stereochemistry. The ion H_2F^+ is predicted to be angular and to have a similar bond angle to water, but its structure is not known. Fluorine bridging is found in a number of compounds, e.g., $(NbF_5)_4$, $(MoF_5)_4$, and $(SbF_5)_n$. The M–F–M angles in these compounds are quite large and may even reach 180°. It would appear that the two bonding electron pairs tend to take up positions

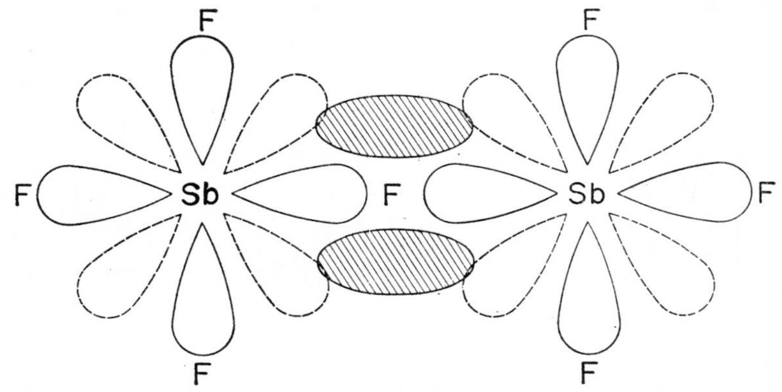

FIG. 6.17 Cross section through a fluorine bridge in an $(SbF_5)_n$ polymer showing delocalization of the fluorine lone-pairs (shown shaded) into the vacant space (d-orbitals) in the valence shell of the antimony (shown in dotted lines).

at 180° with respect to each other, which will tend to force the lone-pairs on fluorine away from the nucleus into the secondary valence shell. This would not be a stable arrangement in the absence of suitable ligands, but it would seem that this arrangement is stabilized in the presence of heavy metals such as antimony or molybdenum which are strong electron pair acceptors and have vacant space (orbitals) in their valence shells in close proximity to the fluorine lone-pairs as illustrated in Fig. 6.17.

In the solid state HF has a hydrogen-bonded polymeric chain structure with a bond angle at fluorine of 120°. The zig-zag nature of this chain presumably arises from the tetrahedral arrangement of fluorine electron pairs although the bond angle of 120° is unexpectedly large (Fig. 6.18).

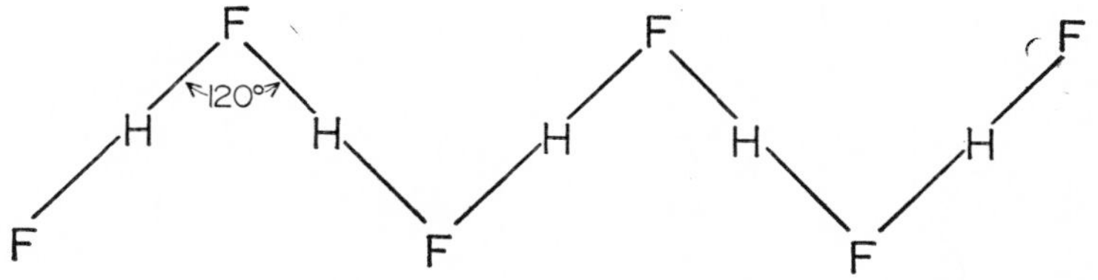

FIG. 6.18 The structure of HF in the solid state.

REFERENCES AND SUGGESTIONS FOR FURTHER READING

Tables of Interatomic Distances and Configurations in Molecules and Ions, Special Publication No. 11, The Chemical Society, London, 1958.

Supplement, Special Publication No. 18, The Chemical Society, London, 1965.

F. A. COTTON and G. WILKINSON, *Advanced Inorganic Chemistry*, 2nd Ed., Interscience, 1966.

R. J. GILLESPIE, *J. Amer. Chem. Soc.*, **82,** 5978, 1960.

L. PAULING, *Nature of the Chemical Bond*, 3rd Ed., Cornell University Press, 1960.

A. F. WELLS, *Structural Inorganic Chemistry*, 3rd Ed., Oxford University Press, 1962.

7

Molecular Geometry of the Elements of the Third and Subsequent Periods

7.1 THE ALKALI AND ALKALINE EARTH METALS

These elements have large central cores and can therefore, in principle, accommodate a large number of electron pairs in their valence shells. However, since their charge is small they have little tendency to attract much electron density into their valence shells and their compounds may be described as predominantly ionic. Since, in any case, they never have lone-pairs, the stereochemistry of ligands around an alkali metal core is independent of the degree of covalency. In infinite lattices, tetrahedral four co-ordination, octahedral six co-ordination, and cubic eight co-ordination are very commonly found. There are few discrete complexes that have had their structures determined. Spectroscopic data indicates that the fairly stable complex $Na(NH_3)_4^+$ is tetrahedral. All the alkali metals form chelate complexes, e.g., with acetyl acetone, salicylaldehyde and diglyme. These complexes appear to be either four or six co-ordinated and would be expected to have tetrahedral or octahedral structures.

The alkaline earth AX_2 molecules are predicted to have a linear structure and it has in fact been generally assumed that this is the case—although there has been very little experimental evidence. A comprehensive electron diffraction investigation in 1957 indicated that all the gaseous alkaline earth dihalides are linear, although the error in the 180° bond angle given for the calcium, strontium, and barium halides was quite large. Recently electric quadrupole deflection of molecular beams has been used to detect molecules possessing permanent dipole moments. Several of the alkali metal dihalides were found to have permanent dipole moments which can only mean that they are bent and not linear. The results of these experiments are summarized in Table 7.1. The linear form is favoured for the lighter central atoms and heavier halogens, while the bent form is favoured for the heavier central atoms and the lighter halogens. The outermost layer of the core of these heavy alkali metals is not very densely filled and can, without too much difficulty, accommodate additional electron density. Indeed, the following element in the periodic table is a transition element in which a further electron has been added to this inner shell. Moreover, the electrons in this shell are not too tightly held and are rather polarizable. It is not surprising therefore that when the outer electron pairs have been reduced in size by combination with an electronegative halogen they may be able to, at least partially, enter the outer shell of the core. In this case this outer shell can no longer be regarded as spherical, and one must consider

Table 7.1 Geometry of the alkaline earth dihalides

	F	Cl	Br	I
Be	l	l	l	l
Mg	l	l	l	l
Ca	b	l	l	l
Sr	b	b	l	l
Ba	b	b	b	b

b = bent; l = linear.

the most probable relative locations of the four electron pairs in this shell and their possible interactions with the two bonding pairs. The four electron pairs in the outer shell of the core will have a relative tetrahedral arrangement, and it would seem reasonable that in order to minimize their interactions with these electrons the two bonding pairs would insert themselves into this shell, one in each of two faces of this tetrahedron, thus giving a bond angle of 109·5° (Fig. 7.1). Depending on how much interaction there is between the bonding electrons and the core electrons a bond angle of between 180° and 109·5° would be expected for the alkali metal dihalides.

These elements all form a number of rather weak complexes, e.g., $MgCl_4^{2-}$, $Mg(acetylacetone)_2$, $Ca(EDTA)^{2-}$ and $Mg(NH_3)_6^{2+}$ which, presumably, have tetrahedral and octahedral structures. The expected tetrahedral stereochemistry has been established in the case of the Grignard reagent $C_6H_5MgBr.2(C_2H_5)_2O$. Diethyl magnesium has a polymeric structure similar to $Be(CH_3)_2$ (Fig. 6.4).

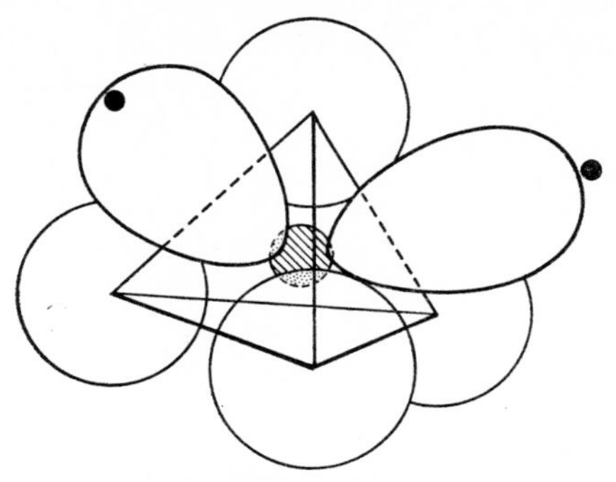

FIG. 7.1 Interaction of two bonding electron pairs, e.g., in BaF_2 with the incomplete outer shell of the core.

7.2 ALUMINIUM, GALLIUM, INDIUM, AND THALLIUM

Since these elements have three electrons in their valence shells they can, in principle, use all three for forming bonds giving the +3 oxidation state, or only one, leaving one unshared pair of electrons, and thus giving the +1 oxidation state. Since they have large valence shells capable of accommodating at least nine electron pairs this places no limitation on the co-ordination numbers that they may reach, which are limited however by the amount of charge that they are able to accept. Thus a co-ordination number of six is only reached with electronegative ligands such as fluorine. The observed molecular geometries for these elements are summarized in Table 7.2.

Table 7.2 Observed molecular geometries for aluminium, gallium, indium, and thallium

Number of lone-pairs and bonds	Arrangement	Lone-pairs	Bonds	Geometry		Example
2	Linear	0	2	AX_2	linear	$Tl(CH_3)_2^+$
3	Planar triangular	0	3	AX_3	planar triangular	$AlCl_3$
4	Tetrahedral	1	3	AX_3E	trigonal pyramidal	$(TlOEt)_4$
		0	4	AX_4	tetrahedral	$AlCl_4^-$
5	Trigonal bipyramid	0	5	AX_5	trigonal bipyramid	$InCl_3(Ph_3P)_2$
6	Octahedral	0	6	AX_6	octahedral	AlF_6^{3-}

(a) AX_2 Co-ordination: Linear Geometry

The $Tl(CH_3)_2^+$ ion has the expected linear structure.

(b) AX_3 Co-ordination: Trigonal Planar Geometry

These elements might be expected to form planar three-co-ordinated molecules like the boron halides. However, there is an extremely strong tendency for them to attract additional electron pairs into their incomplete larger valence shells and therefore they exist mainly as tetrahedral and octahedral complexes. Triphenyl gallium and triphenyl indium exist as planar trigonal molecules apparently linked together by weak intermolecular metal–carbon bonds giving trigonal bipyramidal co-ordination around each metal atom. The molecule $Al[N(Si(CH_3)_3)_2]_3$ has the expected trigonal planar structure and there is good spectroscopic evidence for the existence of planar AX_3 molecules in the vapours of aluminium, gallium and indium chlorides, bromides, and iodides at high temperatures. There is some evidence that gallium triiodide exists as a monomer that is presumably planar and triangular. It is of interest to speculate that thallium and indium trifluoride and possibly the chlorides might have pyramidal rather than planar structures for the same reason as have been given to account for the angular structure of gaseous BaF_2.

It has been stated on the basis of spectroscopic measurements that Me_2TlPy^+ has a T-shaped structure. This is unexpected, as a nearly regular planar trigonal structure would be predicted.

(c) AX_3E Co-ordination: Trigonal Pyramidal Geometry

Although a few Ga(I) and In(I) compounds are known there is no information on their stereochemistry. For thallium however a large number of stable compounds are known containing thallium in the +1 oxidation state. These compounds containing the Tl^+ cation are predominantly ionic, and the only feature of their structures that is of particular interest is that the unshared pair of electrons generally appears to have no influence on the arrangement of ligands around the thallium, and it must therefore be assumed to occupy a spherical orbital. It is said to be a stereochemically inert lone-pair. Such stereochemically inert pairs are sometimes found for elements with very large cores as a single electron pair surrounding such a core in a spherical orbital is considerably delocalized and therefore correspondingly stabilized. The tetrameric thallium ethoxide is an interesting compound in which each thallium is three-co-ordinated and each oxygen four-co-ordinated (Fig. 7.2). The co-ordination around oxygen is tetrahedral as expected, and that around thallium is pyramidal, the lone-pair, which in this case is stereochemically active, occupying the fourth tetrahedral position.

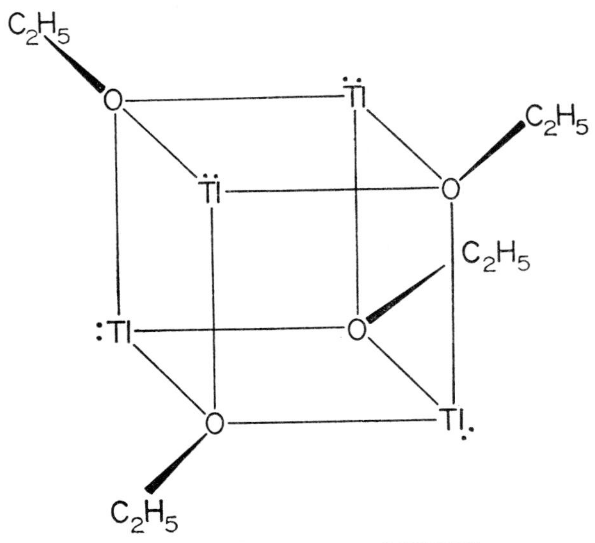

FIG. 7.2 The structure of $(TlOEt)_4$.

(d) AX_4 Co-ordination: Tetrahedral Geometry

The chlorides, bromides, iodides, and alkyls of these elements exist as dimeric molecules with tetrahedral co-ordination around each metal atom, e.g., structures (1), (2), and (3). The angle made by the bridging halogens at the metal is of necessity approximately 90°, and although there may be some bond-bending in these four-membered rings a bond angle of 90° can be tolerated in these large valence shells. The small bond angle at the metal is reflected in the large angle between the terminal halogens. The bond angles in the similar bridged aluminium alkyls (3) are however quite different. This is of course consistent with the fact that these are electron-deficient compounds with only two electron pairs, rather than four, bonding the four bridge atoms together. The bond angles are close to those found for polymeric $Be(CH_3)_2$ which was discussed earlier. At high temperatures the halide dimers dissociate into monomers which presumably have planar trigonal structures. Aluminum *t*-butoxide also forms a similar cyclic dimer with tetrahedral co-ordination around aluminium (4).

The anion $[(C_2H_5)_3AlFAl(C_2H_5)_3]^-$ has tetrahedrally co-ordi-

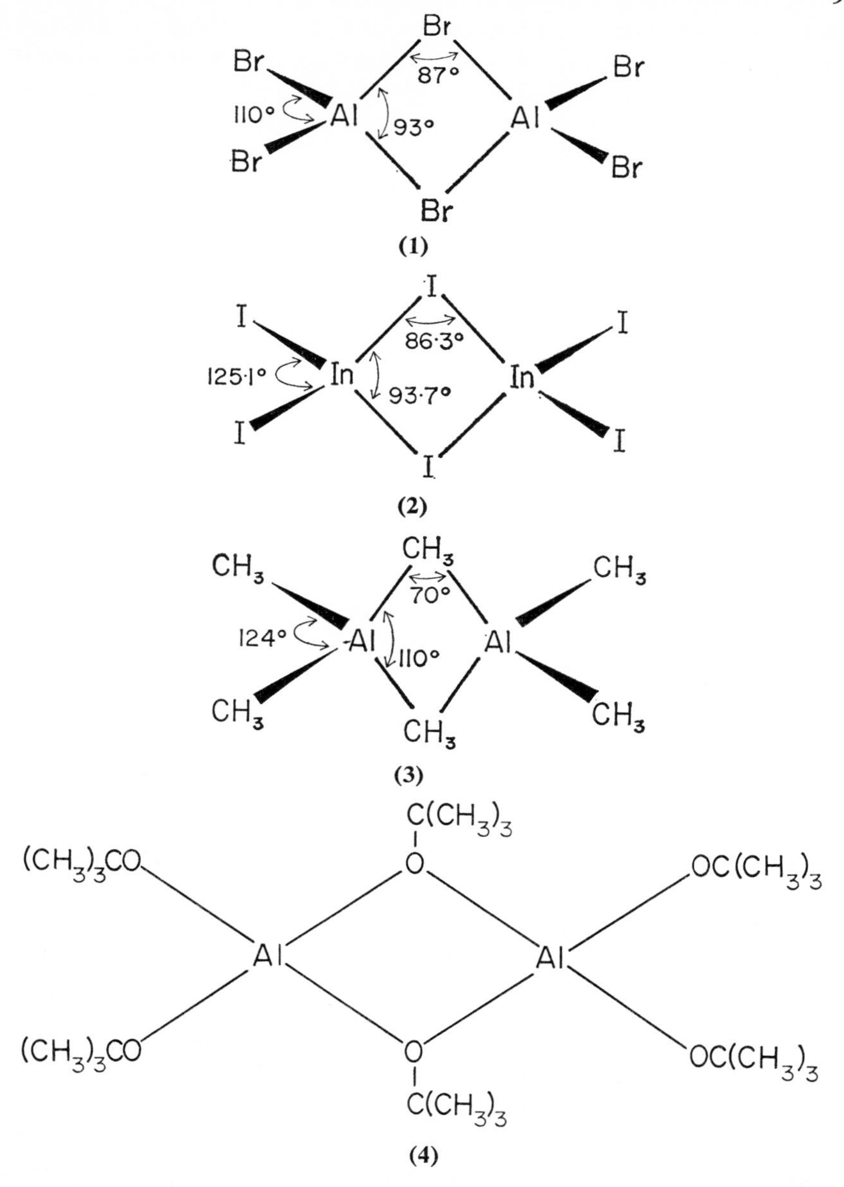

nated aluminium atoms and a linear fluorine bridge. The linear fluorine bridge is quite similar to those found in the tetramer of niobium pentafluoride $(NbF_5)_4$ and the complex aluminofluorides discussed below.

There are many tetrahedral complex anions such as AlH_4^-, GaH_4^-, $AlCl_4^-$, $GaBr_4^-$, $InCl_4^-$, and $Al(C_2H_5)_4^-$ as well as neutral complexes such as $GaH_3.PMe_3$. The aluminium alkyl adducts with primary and secondary amines and phosphines, e.g., $(C_2H_5)_3Al.NH(CH_3)_2$, can eliminate alkane to give the bridged structures (5) and (6) in which both Al and N have tetrahedral co-ordination.

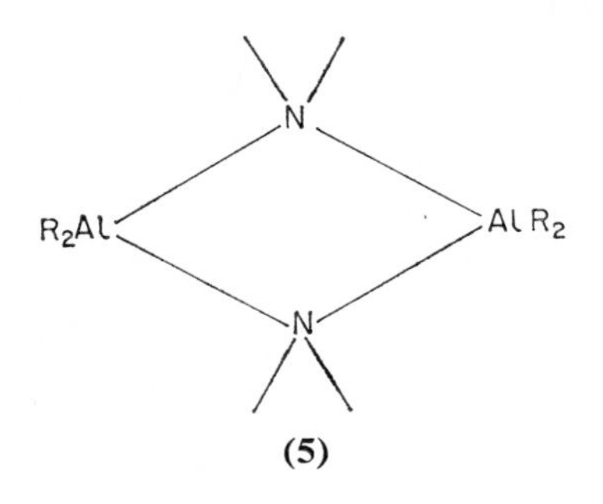

(5)

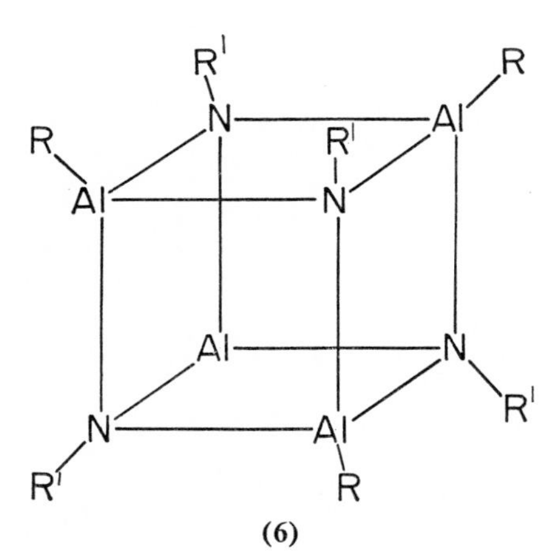

(6)

(e) AX_5 Co-ordination: Trigonal Bipyramidal Geometry

Aluminium trihydride forms several 1 : 2 adducts with amines. The adduct with trimethylamine $AlH_3.2NMe_3$ has the expected trigonal bipyramidal structure with the NMe_3 molecules in the axial positions (7). As the NMe_3 groups are more electronegative than H they are expected to form the longer polar bonds, and they are therefore found in the axial positions in accordance with the arguments given in Chapter 4. The adduct $AlH_3(NMe_2CH_2CH_2NMe_2)$ has a similar structure, in which the nitrogen atoms of one amine molecule

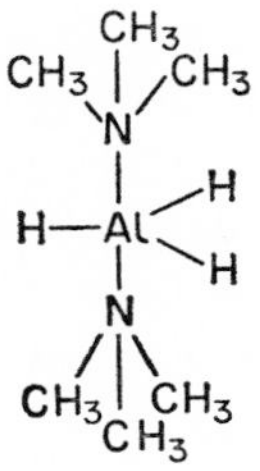

(7)

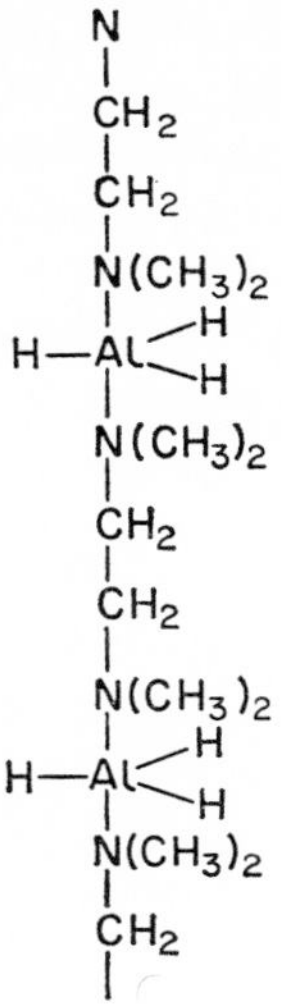

(8)

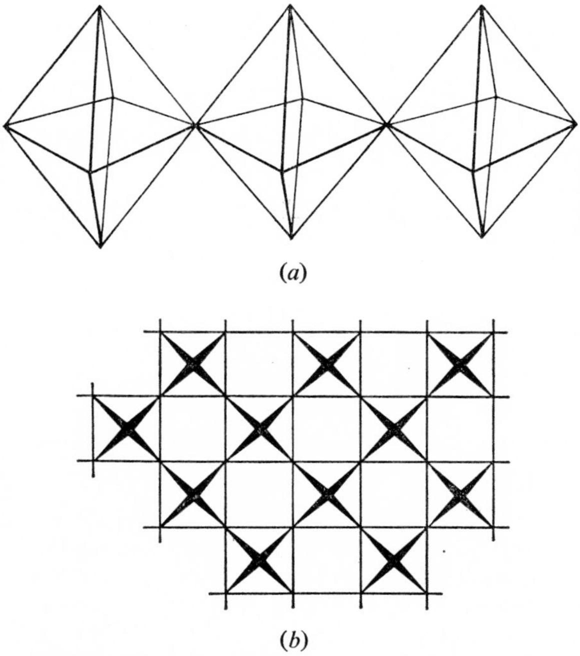

(*a*)

(*b*)

Fig. 7.3 Structures of polymeric fluoroanions of aluminium: (*a*) $(AlF_5^{2-})_n$ (side view); (*b*) $(AlF_4^-)_n$ (top view).

co-ordinate to the axial positions of different aluminium atoms giving a chain structure (8). $InCl_5^{2-}$ unexpectedly has a square pyramid structure but $(Ph_3P)_2InCl_3$ has a trigonal bipyramid structure with axial chlorines.

(f) AX_6 Co-ordination: Octahedral Geometry

These elements form many six co-ordinated molecules. For example the complex ions $Al(H_2O)_6^{3+}$, $Al(OH)_6^{3-}$, and AlF_6^{3-} are all octahedral. Complex fluorides such as Tl_2AlF_5 and $KAlF_4$ also contain octahedral AlF_6 groups sharing two corners to give the linear polymeric $(AlF_5)_n^{2n-}$anion and sharing four corners to give the planar polymeric cation $(AlF_4)_n^{n-}$ respectively (Fig. 7.3) in which each bridging fluorine has linear 180° co-ordination. The binuclear complex anion $Tl_2Cl_9^{3-}$ has a structure just like that of $W_2Cl_9^{3-}$ in which two octahedral $TlCl_6$ groups share a face (Fig. 7.4).

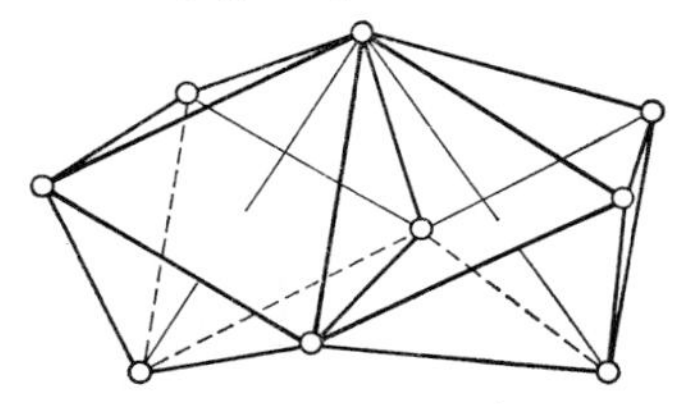

FIG. 7.4 Structure of the $Tl_2Cl_9^{3-}$ anion.

7.3 SILICON, GERMANIUM, TIN, AND LEAD

These elements have four electrons in their valence shells and they can use all four to form bonds giving the +4 oxidation state or they

Table 7.3 Observed molecular geometries for silicon, germanium, tin, and lead

Number of lone-pairs and bonds	Arrangement	Lone-pairs	Bonds	Geometry	Example
3	Triangle	1	2	AX_2E angular	$SnCl_2$
4	Tetrahedron	1	3	AX_3E trigonal pyramidal	$SnCl_3^-$
		0	4	AX_4 tetrahedral	$SiCl_4$
5	Trigonal bipyramid	1	4	AX_4E disphenoidal	
		0	5	AX_5 trigonal bipyramidal	$(CH_3)_3SnBF_4$
6	Octahedron	0	6	AX_6 octahedral	$Pb(OH)_6^{2-}$

may use only two leaving one unshared pair and giving the +2 oxidation state. Their valence shells are large and, even in the case of silicon, can apparently accommodate as many as eight electron pairs, so this places no restriction on the co-ordination numbers that they can reach which are however determined as for the previous group of elements by the amount of charge that they are able to accept. The observed molecular geometries are summarized in Table 7.3.

(a) AX_2E Co-ordination: Angular Geometry

The simple dihalides, $SnCl_2$, $SnBr_2$, SnI_2, $PbCl_2$, $PbBr_2$, and PbI_2 have the expected angular structures in the gas phase, although the angles are not known with any certainty. The salt, $NH_4Pb_2Br_5$, apparently consists of $PbBr_2$ molecules, NH_4^+ and Br^- ions. The $PbBr_2$ molecule is angular with a bond angle of only 85°, but this is perhaps not surprising in view of the large size of the lead core.

(b) AX_3E Co-ordination: Trigonal Pyramidal Geometry

The molecules $SnCl_3^-$ and $SnCl_2.H_2O$ have pyramidal structures with bond angles of slightly less than 90°. GeF_2 has a structure like that of SeO_2 (Fig. 7.24) with one non-bridging fluorine and one bridging fluorine forming an infinite chain and giving pyramidal AX_3E co-ordination around Ge.

(c) AX_4 Co-ordination: Tetrahedral Geometry

The simple molecules of the type MX_4 all have the expected tetrahedral structure, e.g., $SiCl_4$ and $Ge(CH_3)_4$, and compounds with two or more different ligands such as $Sn(CH_3)_2Cl_2$ show the

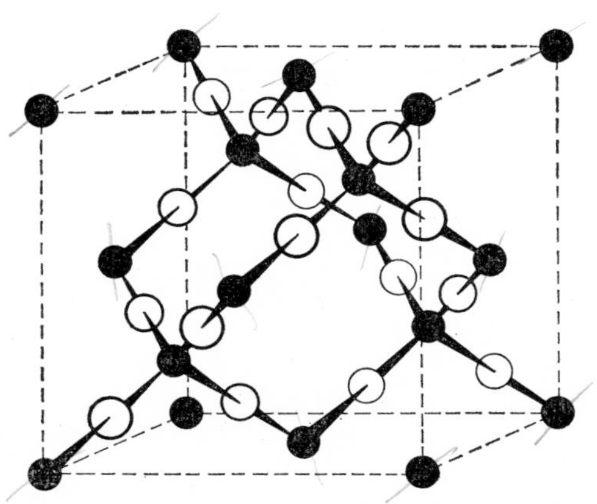

FIG. 7.5 The structure of SiO_2 (β-cristobalite).

expected small deviations from the regular tetrahedral structure. The same tetrahedral arrangement of four bonds is found in the various forms of silica, the silicates, and the siloxanes. For example the β-cristobalite form of silica has the structure shown in Fig. 7.5 with tetrahedral bonds around silicon and a rather large bond angle of 150° at oxygen. The bond angle at an oxygen atom between two silicon atoms varies rather widely from molecule to molecule, and this has been discussed in Chapter 6.

Silicon sulphide contains linear chains which have the structure (9) in which each silicon atom is four co-ordinated and the sulphur atom

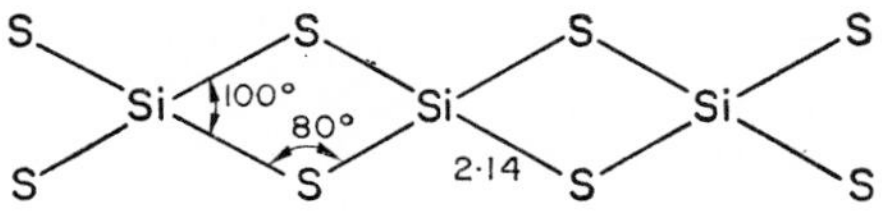

(9)

is two co-ordinated. Germanium sulphide GeS_2 has a similar structure. Because of the four-membered rings thus formed, which presumably have bent bonds, the observed bond angles are expected to be smaller than the angles between the bonding electron pairs. However, it is interesting to note that the larger bond angle of 100° is at silicon while the smaller bond angle of 80° is at sulphur. This is consistent with the presence of lone-pairs on sulphur but not on silicon. If 10° is

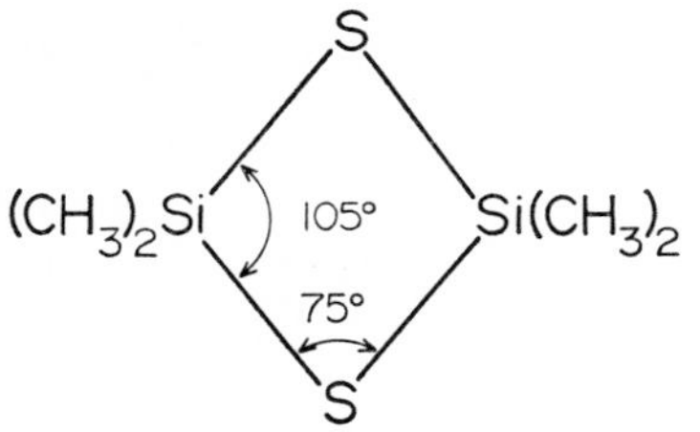

(10)

added to each bond angle to allow for bond-bending then the angle at silicon becomes equal to the tetrahedral angle, and the angle at sulphur becomes 90° which is the smallest angle that is expected to be found at sulphur (p. 57). The cyclic $[Si(CH_3)_2]_2S_2$ molecule has the obviously related structure (10). It is to be noted that the bond angles at a sulphur situated between silicon atoms are always less than tetrahedral and approach 90° as expected for a sulphur atom with an SX_2E_2 configuration, whereas the bond angle at oxygen situated between two silicon atoms is always larger than the tetrahedral angle because of the strong tendency of the lone-pairs to delocalize on to the silicon atoms. Similar considerations apply to GeSGe and GeOGe angles (Table 6.11).

(d) AX_4E Co-ordination: Disphenoidal Geometry

PbO and SnO have rather remarkable structures in which four oxygen atoms lie on one side of a metal atom and equidistant from it (Fig. 7.6). Thus the metal and the four oxygens lie at the corners of a rather flat, square pyramid. The co-ordination around oxygen is

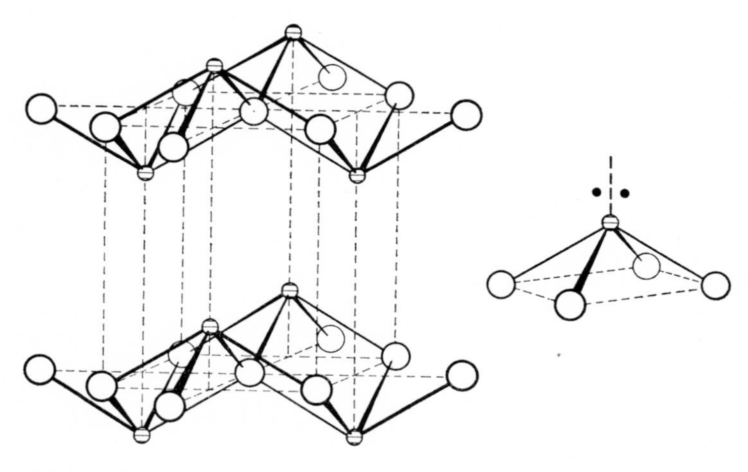

FIG. 7.6 The crystal structure of tetragonal PbO (and SnO). The small shaded circles represent metal atoms. The arrangement of bonds from a metal atom is shown at the right where the two dots represent the unshared pair of electrons.

tetrahedral. The co-ordination around the lead is presumably completed by the lone-pair and the electron-pair arrangement around the lead is therefore a tetragonal pyramid rather than the more usual trigonal bipyramid. But as the difference in energy between these two forms is small, it is quite possible the square pyramid structure could easily be favoured in a more stable three-dimensional structure. PbS has the ionic NaCl structure in which the unshared pair of electrons is stereochemically inactive.

(e) AX_5 Co-ordination: Trigonal bipyramidal Geometry

A number of trimethyltin compounds, e.g., $(CH_3)_3SnBF_4$ and $(CH_3)_3SnClO_4$, have trigonal bipyramidal co-ordination around tin in which planar trimethyltin groups are linked by bridging anions

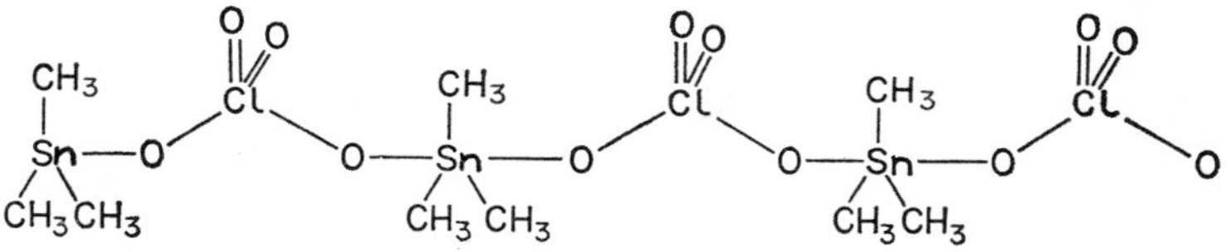

(11)

in the apical positions (11). The fluoride $Sn(CH_3)_3F$ has a similar bridged structure, but the fluorine is situated asymmetrically between the two tin atoms and the structure appears to be intermediate between the symmetrical five-co-ordinated bridged structure and the simple tetrahedral structure for the isolated molecule $(CH_3)_3SnF$(12).

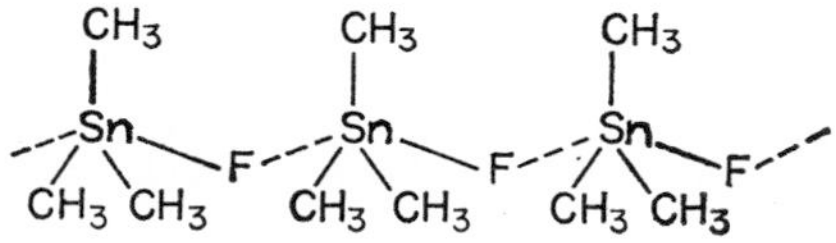

(12)

$SnCl_5^-$ has the expected trigonal bipyramid structure and the SiF_5^- ion which somewhat surprisingly was only recently discovered also has this structure.

(f) AX_6 Co-ordination: Octahedral Geometry

Numerous six-co-ordinated complexes such as SnF_6^{2-}, $Pb(OH)_6^{2-}$, $SnCl_4(Acetone)_2$, $SnCl_4(OPCl_3)_2$, and $GeCl_4.2py$ are known. Octahedral six-co-ordination is also found in one form of GeO_2 and in SnO_2 which have the rutile structure. These are generally described as ionic crystals, but it is reasonable to suppose that the bonds have a certain amount of covalent character. SnF_4 has a polymeric structure like AlF_4^- (Fig. 7.3) with octahedral co-ordination around the tin. SnS_2 has the CdI_2 layer structure in which each tin forms six octahedral bonds and each sulphur has three pyramidal bonds and one lone-pair in the fourth tetrahedral position (Fig. 7.7).

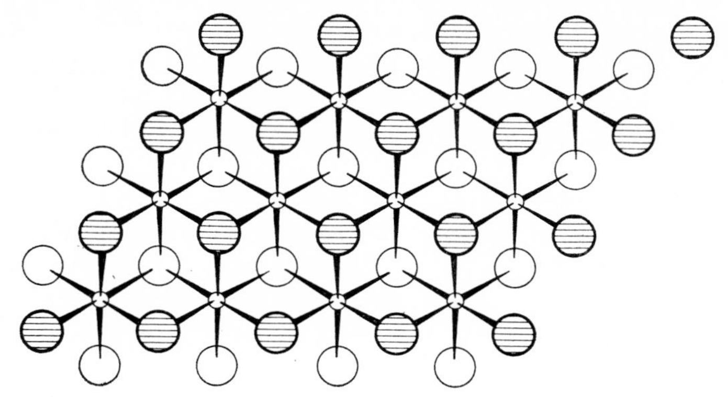

FIG. 7.7 The structure of Sn S_2. Small circles ○ Sn: large circles S atoms, ○ below the plane of the tin atoms; ● above the plane.

7.4 PHOSPHORUS, ARSENIC, ANTIMONY, AND BISMUTH

These elements form compounds either using all their electrons to form bonds, as in the +5 oxidation state, or three of their electrons, leaving one unshared pair as in the +3 oxidation state. The observed molecular geometries are summarized in Table 7.4.

Table 7.4 Observed molecular geometries for phosphorus, arsenic, antimony, and bismuth

Number of lone-pairs and bonds	Arrangement	Lone-pairs	Bonds	Geometry	Example
3	Triangle	1	2	AX_2E angular	PF_2^+
		0	3	AX_3 triangular	$Sb(CH_3)_3^{2+}$
4	Tetrahedron	1	3	AX_3E trigonal pyramidal	$AsCl_3$
		0	4	AX_4 tetrahedral	PCl_4^+
5	Trigonal bipyramid	1	4	AX_4E disphenoidal	$Sb_2F_7^-$
		0	5	AX_5 trigonal bipyramidal	PCl_5^-
6	Octahedron	1	5	AX_5E square pyramidal	$SbCl_5^{2-}$
		0	6	AX_6 octahedral	PCl_6^-
7	Monocapped octahedron	1	6	AX_6E distorted octahedral	$BiBr_6^{3-}$

(a) AX_2E Co-ordination

The compounds $PF_3.AsF_5$, and $PF_3.SbF_5$ have been assigned fluorine bridged ionic structures $PF_2^+MF_6^-$ containing the angular PF_2^+ cation, on the basis of spectroscopic evidence.

(b) AX_3 Co-ordination

The planar $Sb(CH_3)_3^{2+}$ cation appears to be present in aqueous solutions of $Sb(CH_3)_3(ClO_4)_2$, and $Sb(CH_3)_3(NO_3)_2$. In the solid state $Sb(CH_3)_3(NO_3)_2$ is probably trigonal bipyramidal with axial nitrate groups.

(c) AX_3E Co-ordination: Trigonal Pyramidal Geometry

Bond angles for the trihalides and trihydrides are given in Table 7.5.

The lone-pair causes the bond angles to be less than the tetrahedral angle in every case. The bond angles decrease from left to right in the table as the electronegativity of the central atom decreases, and they increase from the fluoride to the iodide with decreasing electronegativity of the ligand. If the lone-pair in PF_3 is co-ordinated with BH_3 its repulsive effect is decreased and the mutual repulsion of the PF bonds causes the bond angle to increase slightly to 99·8° (13). The bond lengths also decrease slightly because of the

Table 7.5 Bond angles for the trihydrides and trihalides

	N	P	As	Sb
H	107·8°	93·3°	91·8°	91·3°
F	102·1	97·8	96·2	88
Cl		100·3	98·5	99·5
Br		101·5	99·6	97
I		102	100·2	99·1

increased effective electronegativity of phosphorus when it acquires a positive charge. Alternatively we may note that as the lone-pair is effectively anti-bonding it weakens the P—F bonds somewhat in PF_3. This antibonding effect is lost when the lone pair co-ordinates with BH_3 and the P—F bonds are correspondingly strengthened.

The angles in the hydrides also decrease in the expected manner with decreasing electronegativity of the central atom from phosphorus to antimony. However the bond angles in the hydrides are generally smaller than in the fluorides despite the greater electronegativity of fluorine which should lead to the repulsions between the bonding pairs being smallest in the case of fluorine. It seems that the bond angles in hydrides of the heavier elements are generally anomalously small when compared with the angles between other ligands. This can be attributed to the unique character of hydrogen as a ligand. Thus the density of the bonding electron pair is located not only in the bonding region but it is also spread out around the hydrogen nucleus, whereas for all other ligands a large amount of the space in the valence shell of the ligand is taken up by other electron

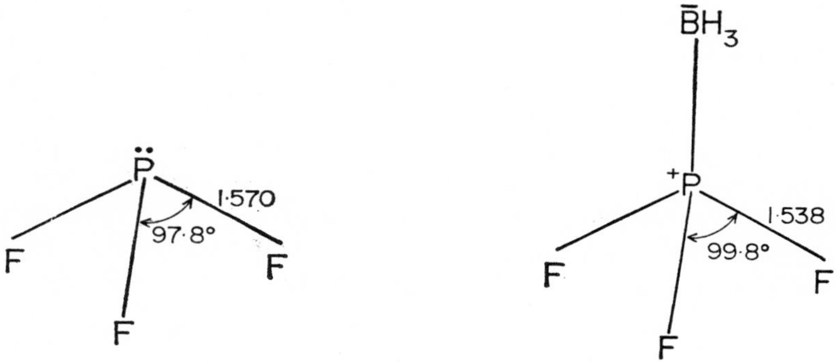

(13)

pairs so that the electron density of the bonding pair is more concentrated in the bonding region. We may summarize then, by saying that the amount of electron density in an X—H bond, and hence the size of the bonding orbital, are less than would be expected from electronegativity considerations alone.

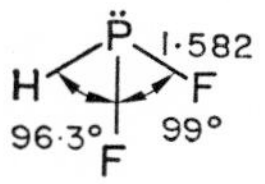

(14)

In agreement with this we note that when one fluorine is replaced by hydrogen to give PHF_2 (14), the bond lengths increase as expected because the lower electronegativity of hydrogen decreases the effective electronegativity of the phosphorus, but at the same time the bond angle decreases because of the smaller size of the P—H bonding pair. We note also that because the phosphorus in HPF_2 has a greater effective electronegativity than in PH_3, the PH bond length is shorter in PHF_2 than in PH_3. The bond angles for some other AX_3E molecules are given in Table 7.6. The angles in the two cyanides are particularly small. $P(SiH_3)_3$ and $As(SiH_3)_3$ are pyramidal, unlike $N(SiH_3)_3$ which is planar. In contrast to nitrogen, phosphorus and arsenic, which have incomplete valence shells in these molecules, have little tendency to delocalize their lone-pairs.

Table 7.6 Bond angles for some AX_3E molecules

	P	As	Sb
$M(CN)_3$	93°	92°	
$M(CH_3)_3$	99·1	96	
$M(CF_3)_3$	99·6	100·1	100·0°

P_2I_4 has the centrosymmetrical *trans* structure shown in Fig. 7.8 with a pyramidal configuration around each P atom. The IPI angle is 102°, the same as in PI_3.

The tetrahedral P_4 molecule has an AX_3E configuration around each phosphorus, and although the bond angles are only 60° it is reasonable to presume that there is considerable bond bending due to bond–bond repulsions, and that the maximum of the electron density in each bond lies outside the internuclear P–P axis (Fig. 7.9).

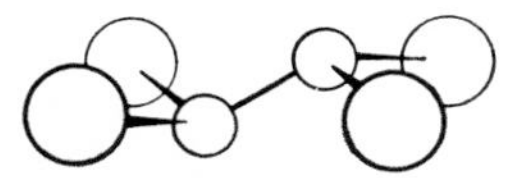

FIG. 7.8 The structure of the P_2I_4 molecule.

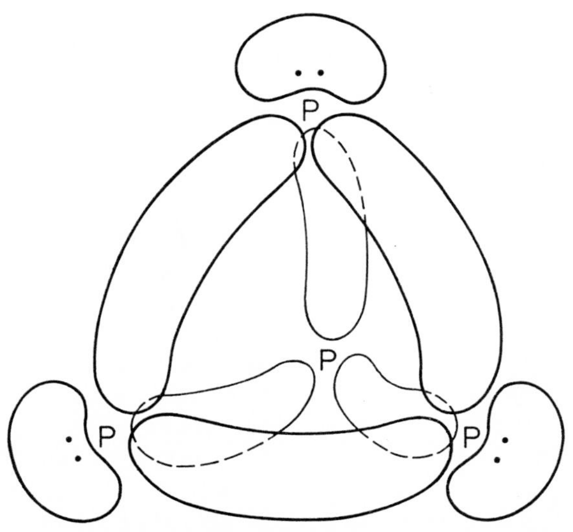

FIG. 7.9 Bonding and lone-pair orbitals in the P_4 molecule.

Other molecules in which the co-ordination is of the AX_3E type include the oxides of the +3 oxidation state of these elements. These are of several different types. The molecules P_4O_6, As_4O_6, and Sb_4O_6 have the structure shown in Fig. 7.10. Another form of arsenic (III) oxide has a layer structure with bond angles at arsenic of approximately 100° (Fig. 7.11) and antimony (III) oxide has a double chain structure with bond angles at antimony of 81°, 93°, and 99° and large bond angles of 116° and 132° at oxygen (Fig. 7.12). The poly meta-arsenite ion $(AsO_2)_n^{n-}$ consists of a chain of pyramidal AsO_3 groups (Fig. 7.13).

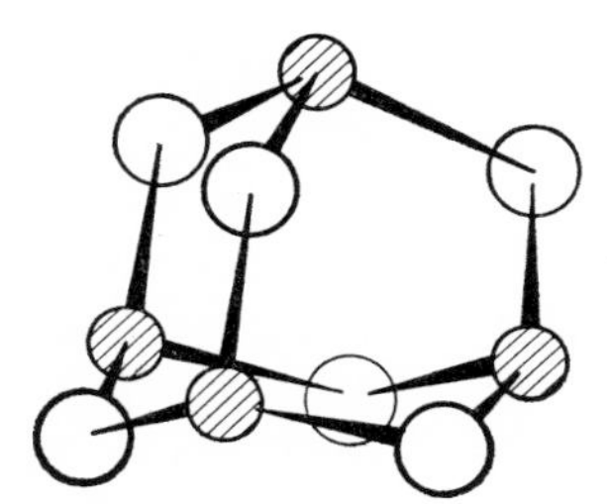

FIG. 7.10 The structure of the P_4O_6, As_4O_6, and Sb_4O_6 molecules. P, As or Sb shaded circles, O open circles.

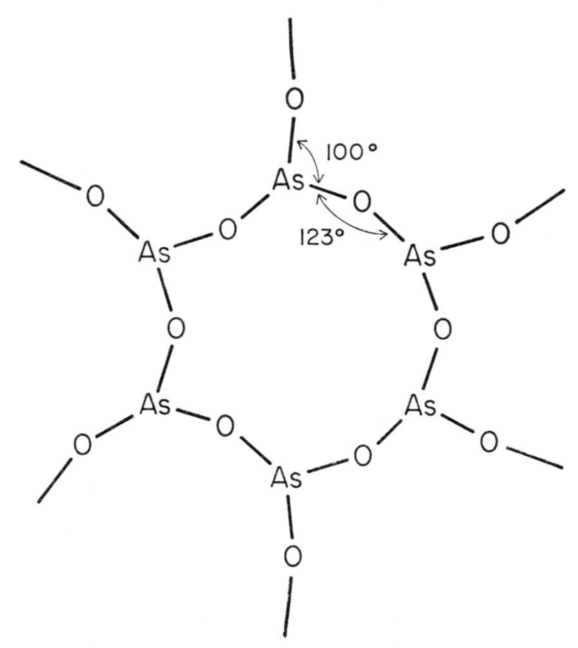

FIG. 7.11 Layer structure of arsenic(III) oxide.

The phosphorus sulphide P_4S_3 also contains four pyramidal phosphorus atoms (Fig. 7.14). The bond angles in the P_3 ring are of course only 60°, but as usual, we presume that there is a considerable amount of bond bending as in P_4. P_4S_5, P_4S_7 and $P_4S_3I_2$ also have similar structures (Fig. 7.14). As_4S_4 has an interesting related structure containing pyramidal arsenic atoms with a bond angle at arsenic

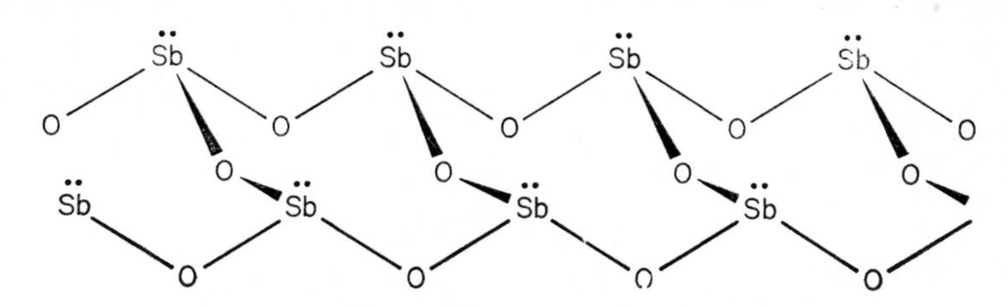

FIG. 7.12 Chain structure of antimony(III) oxide.

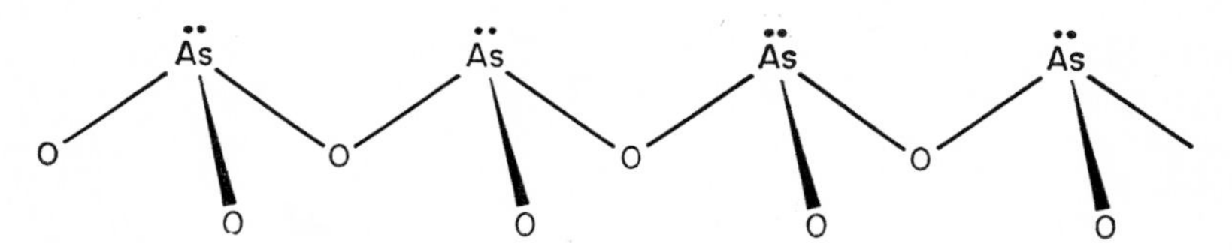

FIG. 7.13 The polymeta-arsenite ion $(AsO^-)_n$.

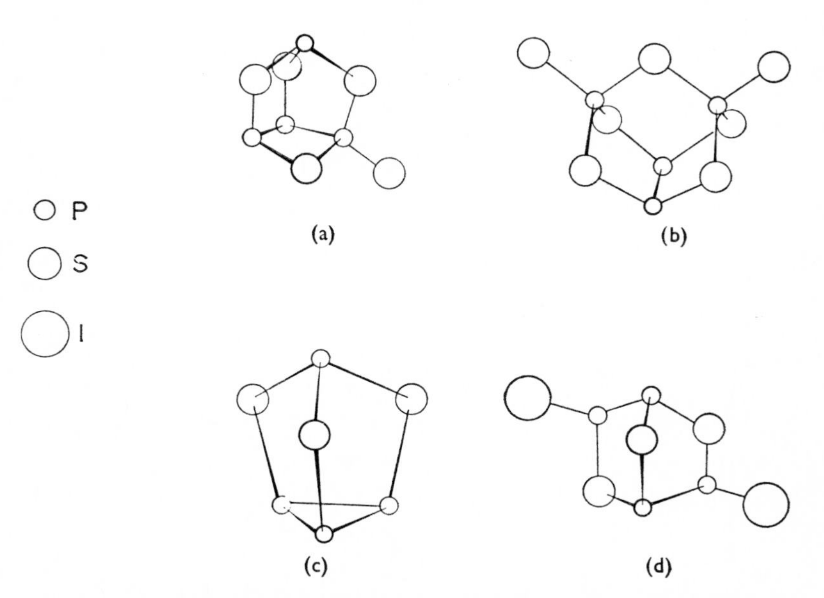

FIG. 7.14 The structures of (*a*) P_4S_5; (*b*) P_4S_7; (*c*) P_4S_3; (*d*) $P_4S_3I_2$.

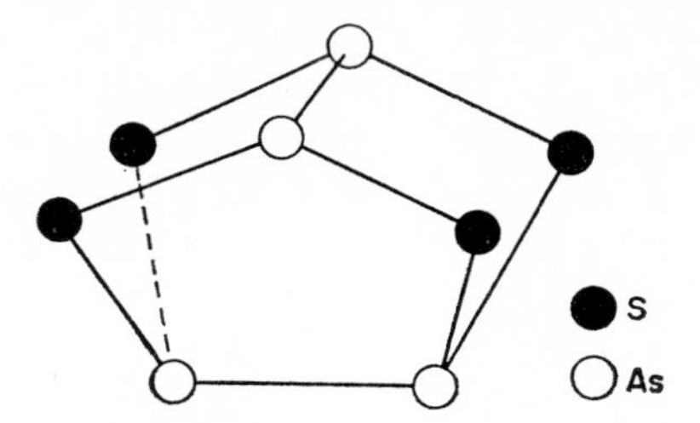

FIG. 7.15 The structure of As_4S_4.

of 93° (Fig. 7.15). Orpiment, As_2S_3, has a layer structure similar to that of As_2O_3.

(d) AX_4 Co-ordination: Tetrahedral Geometry

There are numerous molecules of the type PX_4^+ or O=PX_3 or S=PX_3. They all have the expected tetrahedral shape. When the ligands are identical, e.g., PH_4^+, PCl_4^+, $AsPh_4^+$, and PO_4^{3-}, the tetrahedra are regular. In molecules of the type POX_3 the XPX angles are always found to be less than 109·5° as a consequence of the greater repulsion exerted by a double bond than by a single bond (see Table 7.7). The di- and tri-phosphates and other related condensed anions have a tetrahedral arrangement of bonds around

Table 7.7 Bond lengths and bond angles in POX_3 and PSX_3 molecules

	P—X Å	P=O Å	XPX
POF_3	1·52	1·44	101·3°
$POCl_3$	1·99	1·45	103·3
$POBr_3$	2·06	1·41	108
		P=S	
PSF_3	1·53	1·87	100·3
$PSCl_3$	2·01	1·88	101·8
$PSBr_3$	2·13	1·89	106

phosphorus. Phosphoric oxide P_4O_{10} and the related P_4O_9 and $P_4O_6S_4$ molecules also contain tetrahedrally co-ordinated phosphorus atoms (Fig. 7.16).

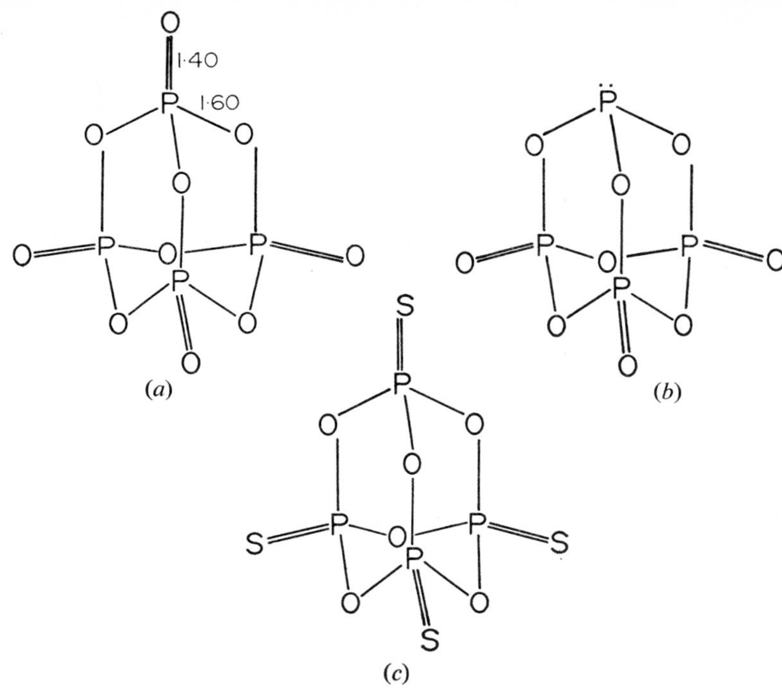

FIG. 7.16 The structures of (*a*) P_4O_{10}; (*b*) P_4O_9; (*c*) $P_4O_6S_4$.

(e) AX_4E Co-ordination: Disphenoidal Geometry

The molecules $SbCl_3.C_6H_5NH_2$ and $Sb(SCH_2CO_2)_2H$ have the expected trigonal bipyramid arrangement of two axial ligands, two equatorial ligands, and an equatorial lone-pair around antimony with the more electronegative ligands occupying the axial positions (Fig. 7.17). In one crystalline form of Sb_2O_4 there are six co-ordinated Sb(V) atoms and four co-ordinated Sb(III) atoms. The latter have the expected disphenoidal geometry, and the distortions of the bond angles and bond lengths produced by the lone-pair are clearly evident (Fig. 7.17).

(f) AX_5 Co-ordination: Trigonal Bipyramidal Geometry

With one exception all the AX_5 molecules of the Group V elements have the expected trigonal bipyramid shape. In all cases where they have been measured the axial bonds are longer than the equatorial bonds, and if there are different ligands the more electronegative ligands occupy the axial positions. Details of these structures have been given in Table 4.1. Other molecules whose structures have

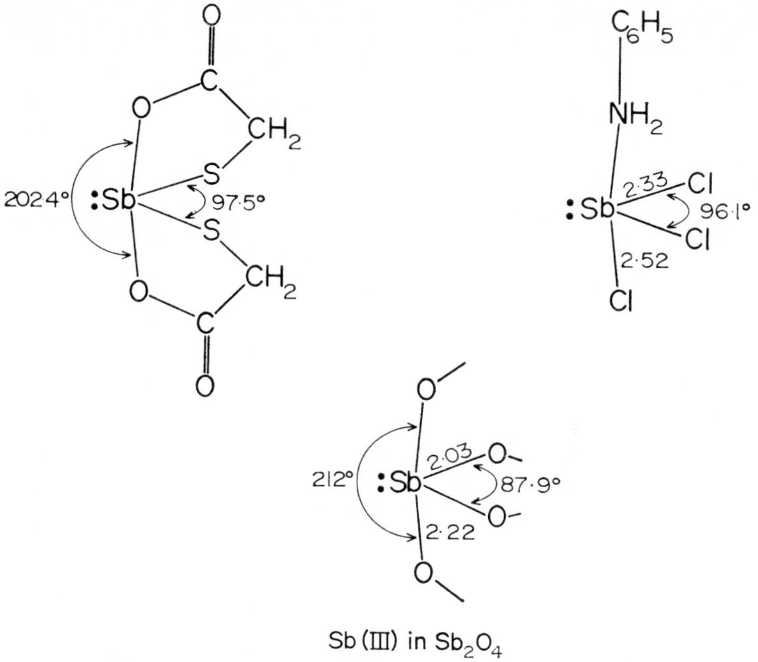

Sb (III) in Sb_2O_4

FIG. 7.17 The structures $SbCl_3.C_6H_5NH_2$, and $Sb(SCH_2CO_2)_2H$ and the geometry of Sb(III) in Sb_2O_4.

been established spectroscopically include PH_2F_3, PF_4Cl, PF_3Cl_2, $PFBr_2$, and $PF_3(CH_3)_2$. In each case the most electronegative ligand, i.e., fluorine, occupies the axial positions. The dimeric molecule

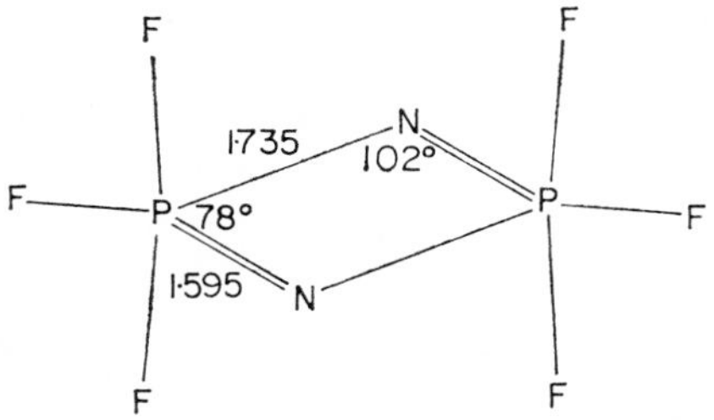

(15)

$(F_3PNCH_3)_2$ has a planar P_2N_2 ring and trigonal bipyramidal co-ordination at phosphorus (15). Because of the smaller size of the nitrogen valency shell it is more difficult to reduce the bond angle from the expected 120° to the value of 90° required by a rectangular ring than it is to reduce the bond angle at phosphorus. Consequently the bond angle at N is considerably larger than at P. Contrary to expectation $Sb(C_6H_5)_5$ has been found to have a square pyramid structure. The only other AX_5 molecule known to have a square pyramid structure is $InCl_5^{2-}$. The reason for this anomalous behaviour is not clear, but the difference in energy between the trigonal bipyramidal and square pyramidal structures is small, and packing considerations in the crystal could cause the square pyramid structure to be adopted. Certainly it would be expected that it would be easier to distort a trigonal bipyramid to a square pyramid the larger is the central atom and hence the smaller the bond–bond repulsions; thus it is reasonable that antimony pentaphenyl should have the square pyramid structure rather than the arsenic or phosphorus compounds.

(g) AX_5E Co-ordination: Square Pyramid Geometry

These molecules have a square pyramid structure. Examples include anions of the general type AX_5^{2-}, e.g., SbF_5^{2-} (16), $SbCl_5^{2-}$

(16) (17) (18)

(17), $BiCl_5^{2-}$ (18), some complex anions and some neutral complexes $AX_3.2L$. The complex ion $Sb_4F_{16}^{4-}$ consists of a ring of four square pyramids joined by linear fluorine bridge bonds (Fig. 7.18). The

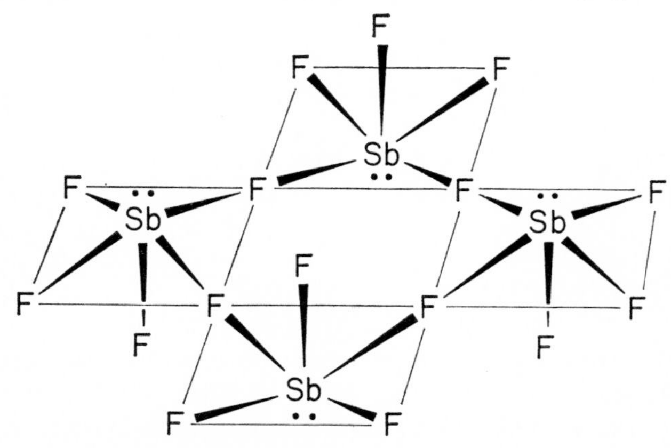

FIG. 7.18 The structure of the $Sb_4F_{16}^{4-}$ ion.

$Bi_2Cl_8^{2-}$ ion has a square pyramid co-ordination around each Bi and two bridging chlorines (19). As is usually the case for molecules of this type, the central atom lies slightly below the plane of the four ligands

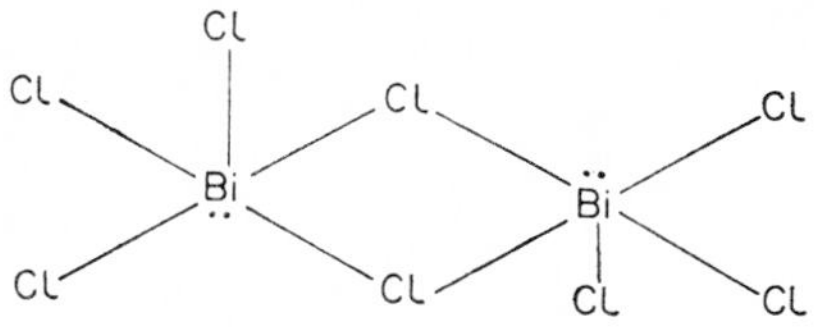

(19)

in the base of the square pyramid because of the repulsion of the bond pairs by the lone-pair.

In $NaSbF_4$ the SbF_4^- ion has a linear polymeric structure with an octahedral co-ordination of five fluorines and a lone-pair around each antimony, each antimony forming two fluorine bridges.

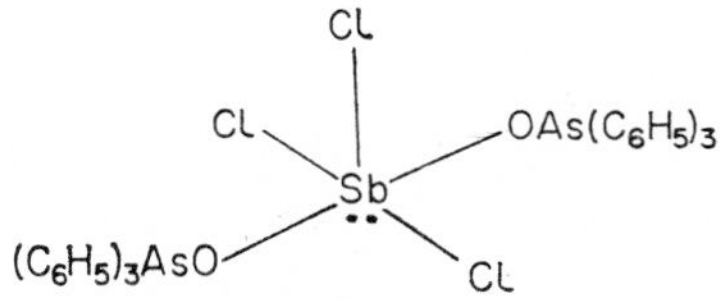

(20)

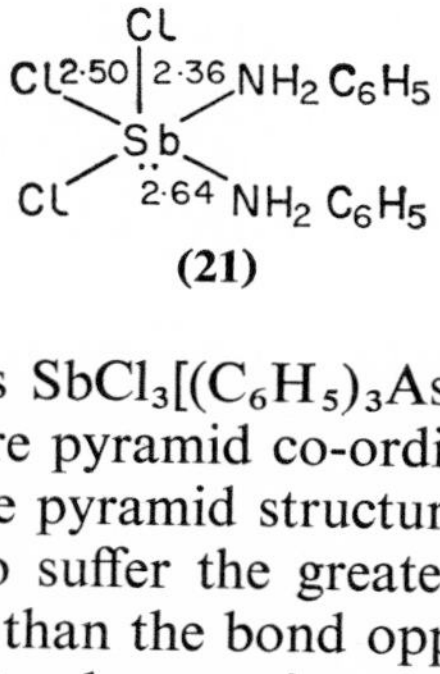

(21)

The neutral complexes $SbCl_3[(C_6H_5)_3AsO]_2$, $SbCl_3(C_6H_5NH_2)_2$, also have the same square pyramid co-ordination around antimony (20), (21). In these square pyramid structures the bonds next to the lone-pair are expected to suffer the greatest repulsion (p. 44, Fig. 3.9) and hence be longer than the bond opposite the lone-pair. This is the case for all the molecules mentioned above except SbF_5^{2-}.

(h) AX_6 Co-ordination: Octahedral Geometry

Molecules exhibiting this co-ordination include anions of the type MX_6^- and neutral complexes of the type $MX_5.L$. There are many examples of octahedral ions such as PF_6^-, PCl_6^-, $Sb(OH)_6^-$, SbF_6^-. The pentahalides are strong Lewis acids and form complexes of the type $AX_5.L$, e.g., $SbCl_5.POCl_3$ (Fig. 7.19) $SbF_5.SO_2$.

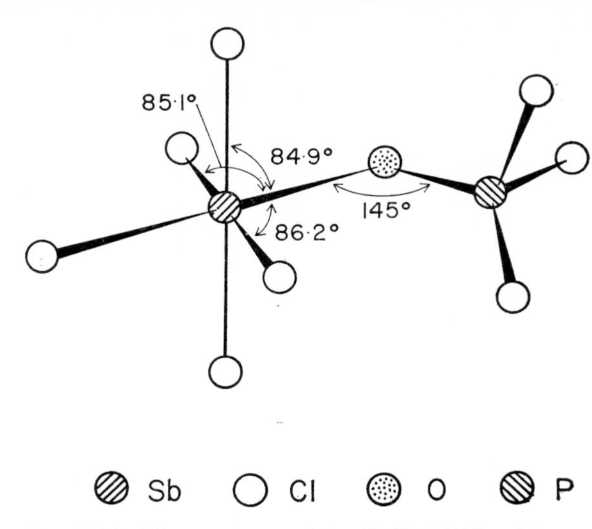

FIG. 7.19 The structure of the $SbCl_5.POCl_3$ adduct.

$SbCl_5.(CH_3)_3PO$, $SbCl_5.Ph_2SO$, $SbCl_5.SeOCl_2$ and $SbCl_5.(CH_3)_2$ which have approximate octrahedral structures. In each case the donor atom is a doubly bonded oxygen atom. On formation of the donor–acceptor complex this oxygen acquires a formal positive charge, and the high electronegativity of this positive oxygen causes the electron-pair bond from antimony to this oxygen to be more contracted and to occupy a smaller orbital than the other bonding

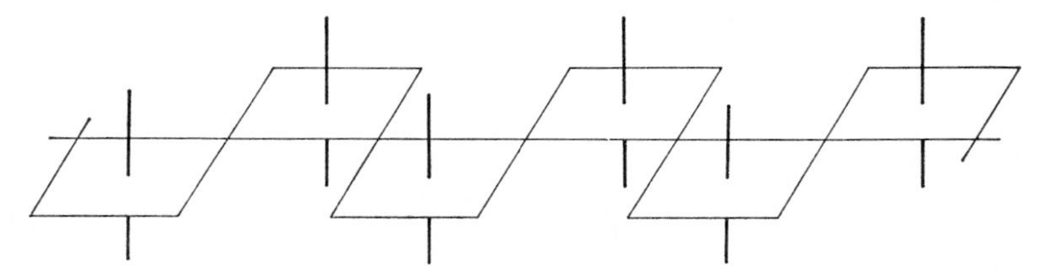

FIG. 7.20 The structure of antimony pentafluoride $(SbF_5)_n$.

electron pairs; and hence the octahedron is distorted, the four neighbouring halogen oxygens being bent slightly towards this oxygen atom.

Antimony pentafluoride has a polymeric structure in the liquid state in which each antimony has an octahedral structure with *cis* bridging to neighbouring SbF_6 groups (Fig. 7.20). Segments of this structure are also found in the $Sb_2F_{11}^-$ and $Sb_3F_{16}^-$ anions (Fig. 7.21). The $Sb_3F_{16}^-$ ion is known to exist in both the *cis* form shown

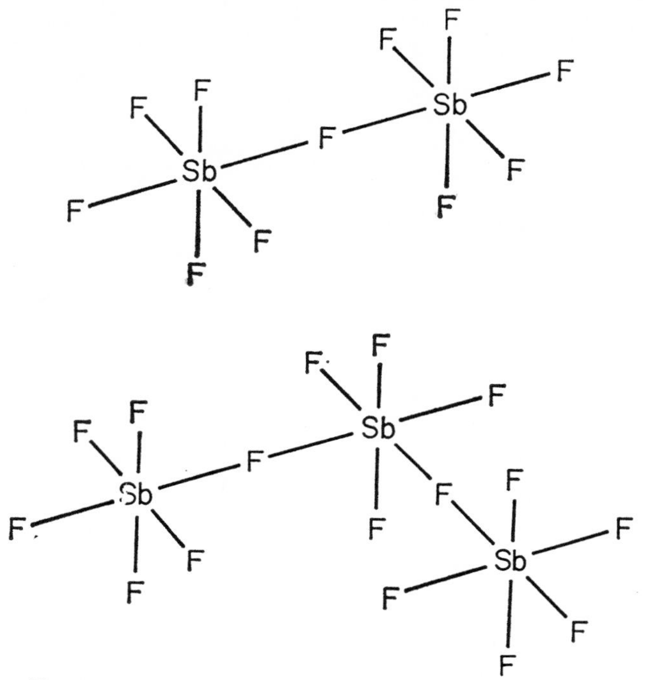

FIG. 7.21 The structures of $Sb_2F_{11}^-$ and *cis* $Sb_3F_{16}^-$.

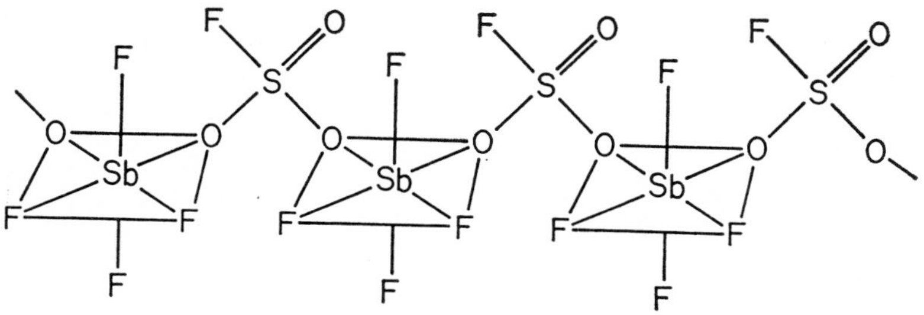

FIG. 7.22 The structure of $(SbF_4SO_3F)_n$.

in Fig. 7.21 and in the *trans* form. An interesting related structure is that of $SbF_4.SO_3F$ in which each antimony acquires octahedral co-ordination by the formation of fluorosulphate bridges (Fig. 7.22).

(i) AX_6E Co-ordination: Distorted Octahedral Geometry

This co-ordination should give a distorted octahedral arrangement of the ligands around the central atom due to the presence of the lone-pair. The exact arrangement of the ligands is difficult to predict because of the various possible alternative arrangements for seven electron pairs but it seems likely that it would be based on the 1 : 3 : 3 arrangement of seven electron pairs with the lone-pair in the unique axial position. The polymeric $BiBr_4^-$, BiI_4^-, and $BiBr_5^{2-}$ ions and the monomeric $BiBr_6^{3-}$ ion all have a distorted octahedral arrangement of six ligands around bismuth, although in the case of the polymeric ions the distortion from octahedral symmetry may be attributed at least partly to the halogen bridging. Raman spectra of solutions strongly indicate that $SbCl_6^{3-}$, $SbBr_6^{3-}$, and SbI_6^{3-} have distorted octahedral structures. However, in the solid state in $(NH_4)_2SbBr_6$ which contains both $SbBr_6^{3-}$ and $SbBr_6^-$ both ions have regular octahedral structures. The isoelectronic TeX_6^{2-} ions also have regular octahedral structures in the solid state and these exceptions to the general rules are further discussed in section 7.5(k).

7.5 SULPHUR, SELENIUM, AND TELLURIUM

This group show a great variety of molecular geometries which are a consequence of the number of stable oxidation states and the possibility of every co-ordination number up to six. The various observed geometries are summarized in Table 7.8.

Table 7.8 Observed molecular geometries for sulphur, selenium, and tellurium

Number of lone-pairs and bonds	Arrangement	Lone-pairs	Bonds	Geometry	Example
3	Equilateral triangle	1	2	AX_2E angular	SO_2
		0	3	AX_3 triangular	SO_3
4	Tetrahedron	2	2	AX_2E_2 angular	$TeBr_2$
		1	3	AX_3E trigonal pyramidal	SOF_2
		0	4	AX_4 tetrahedral	SO_4^{2-}
5	Trigonal bipyramid	1	4	AX_4E disphenoidal	SF_4
		0	5	AX_5 trigonal bipyramidal	SOF_4
6	Octahedron	1	5	AXE_5 square pyramidal	TeF_5^-
		0	6	AX_6 octahedral	SF_6
7	Monocapped octahedron	1	6	AX_6E distorted octahedral*	$SeCl_6^{2-}$

* This is the predicted shape, however, the observed geometry has been found to be regular octahedral in several cases.

(a) AX_2E Co-ordination: Angular Geometry

Sulphur dioxide is an angular molecule with a bond angle of 119·5° which is very close to the ideal angle of 120°. It would appear that the repulsions exerted by the S=O double bonds and the lone-pair have very similar magnitudes. S_2O has a very similar structure with a bond angle of 118°.

(b) AX_3 Co-ordination: Trigonal Planar Geometry

Sulphur trioxide SO_3 and, presumably, also monomeric selenium trioxide SeO_3 have planar triangular structures with SO and SeO double bonds.

(c) AX_2E_2 Co-ordination: Angular Geometry

Bond angles for some simple AX_2E_2 molecules are given in Table 7.9. In general, these are smaller than the tetrahedral angle, and are

Table 7.9 Bond angles in AX_2 molecules of sulphur, selenium, and tellurium

H_2S	92·2°	H_2Se	91°	H_2Te	89·5°
SCl_2	98			$TeBr_2$	98
$S(CH_3)_2$	105	$Se(CH_3)_2$	98		
$S(CF_3)_2$	105·6	$Se(CF_3)_2$	104		
$SH(CH_3)$	99·4				

always smaller than the corresponding angle at oxygen. This is consistent with the greater size and smaller electronegativity of sulphur. In H_2S the two bonding pairs of electrons are further apart than in H_2O because of the greater size of the inner core of sulphur, and also because the smaller electronegativity of sulphur relative to oxygen allows the two electron pairs to move out towards the hydrogens, i.e., the bonds are less polar. The greater repulsions between the two lone-pairs than between the bond pairs causes the bond pairs to move together and no appreciable overlap, and therefore no repulsion, occurs until a considerably smaller angle than in the corresponding oxygen molecule is reached. The approach of the angle to 90° is consistent with our earlier conclusion that the valence shell of sulphur can accommodate at most six electron pairs, and also with the idea that the repulsions between X—H bonds are smaller than expected from the electronegativity of hydrogen (p. 58). The slightly smaller bond angle of 91·0° in H_2Se can be attributed to the smaller electronegativity of selenium. It is significant that the bond angle in H_2Te is very slightly less than 90° as this is consistent with Te being able to

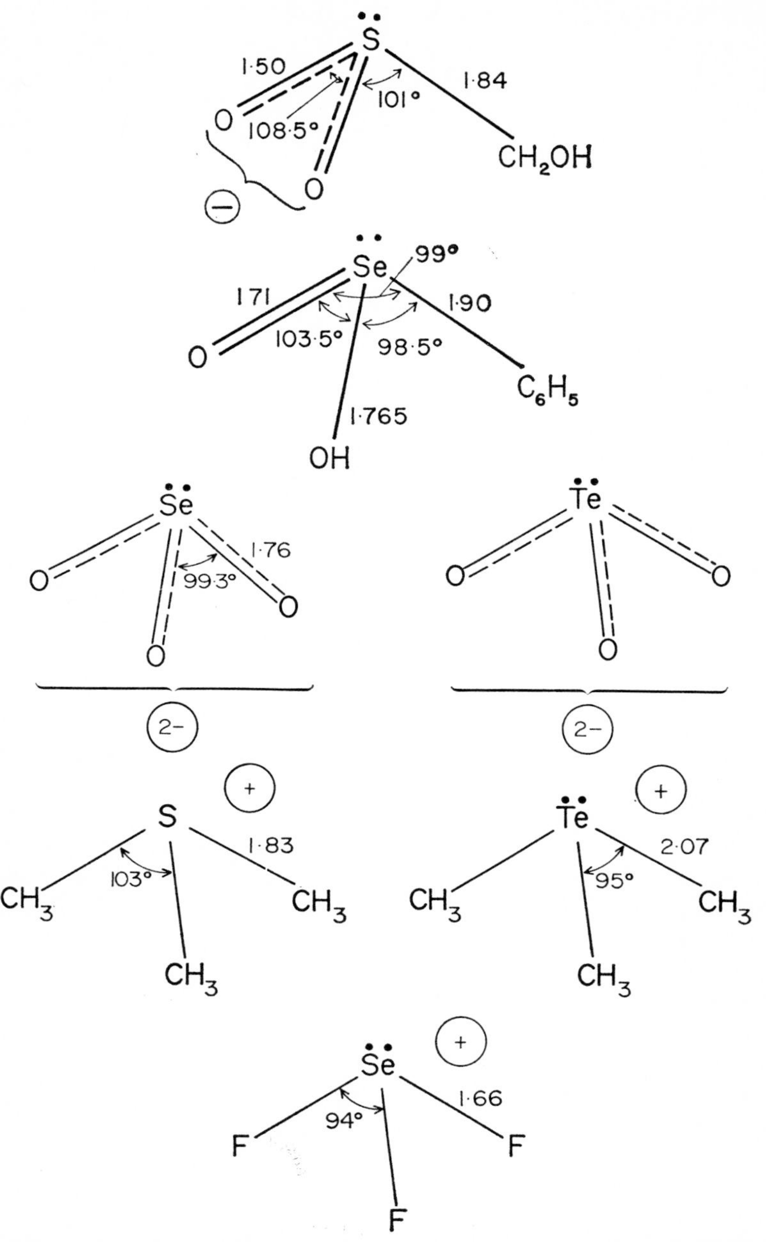

FIG. 7.23 The structures of some trigonal pyramidal AX_3E molecules of sulphur, selenium and tellurium.

accommodate more than six electron pairs in its valence shell which corresponds to an angle of less than 90° if two bonding pairs are pushed as close as possible together. The bond angles in other SX_2 compounds are larger, but nevertheless smaller than the angle at oxygen in the corresponding OX_2 molecule. A number of molecules of the type XS—SX have angular geometry at each sulphur atom. Typical examples are 22(a) and 22(b).

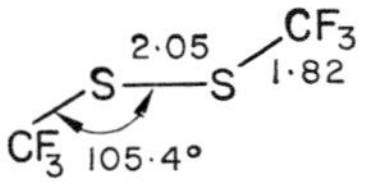

2·05 CF_3 S—S 1·82 CF_3 105·4°

(22a)

2·05 Cl 1·44 S—S Cl 103°

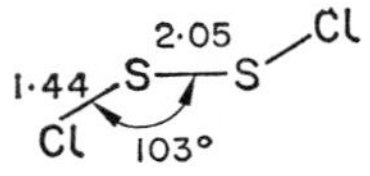

(22b)

(d) AX_3E Co-ordination: Pyramidal Geometry

Molecules having this geometry include AX_3^+ cations, AX_2O and AXO_2 molecules, and AO_3^{2-} axions. They all have the expected pyramidal shape resulting from a tetrahedral arrangement of three

Table 7.10 Bond angles for pyramidal $X_2S(Se)O$ molecules

	XSO	XSX
F_2SO	106·8°	92·8°
Br_2SO	108	96
$(CH_3)_2SO$	107	100
$(C_6H_5)_2SO$	106·2	97·3
$SeOF_2$	104·8	92·2

bonds and one lone-pair. Some structural data is given in Table 7.10 and in Figure 7.23. Selenium dioxide has a polymeric structure with a pyramidal arrangement of oxygens around each selenium (Fig. 7.24) and the selenium oxyhalides $SeOCl_2$ and $SeOBr_2$ have vibrational spectra that are consistent with the expected pyramidal structure.

There are several complex oxyions of sulphur containing a sulphur–sulphur bond, e.g., dithionite $S_2O_4^{2-}$, disulphite $S_2O_5^{2-}$, and dithionate $S_2O_6^{2-}$. These have the expected stereochemistry around each sulphur as shown in Fig. 7.25. An unusual feature of these molecules is however that the S—S bond lengths have the values

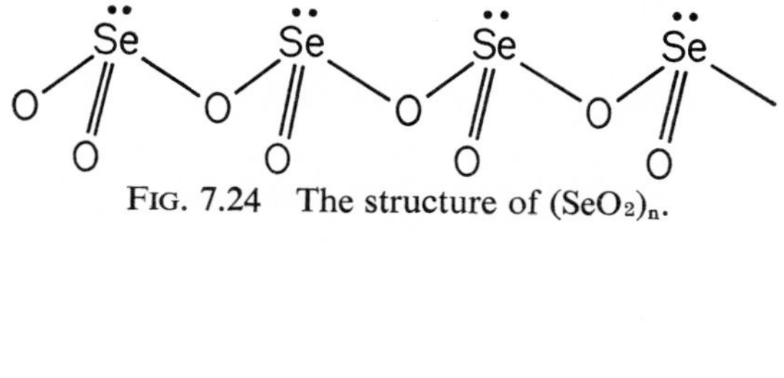

Se Se Se Se O O O O O O O O

FIG. 7.24 The structure of $(SeO_2)_n$.

2·39 Å in $S_2O_4^{2-}$, 2·21 Å in $S_2O_5^{2-}$, and 2·15 Å in $S_2O_6^{2-}$. These bond lengths are all considerably longer than is usually found for S—S bonds (the sum of the covalent radii = 2·08 Å) and are particularly long in the cases of $S_2O_4^{2-}$ and $S_2O_5^{2-}$. The molecule $S_2O_4^{2-}$ is rather similar to the N_2O_4 molecule where the long N—N bond can be regarded as being formed from an electron in the NO_2 molecule that is in the secondary valence shell of nitrogen (p. 115). The valence shell of sulphur can barely accommodate six electron pairs,

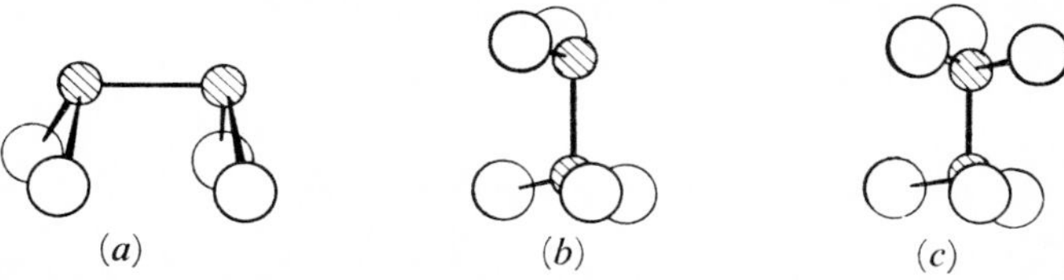

FIG. 7.25 The structures of (*a*) dithionite $S_2O_4^{2-}$; (*b*) disulphite $S_2O_5^{2-}$, and (*c*) dithionate $S_2O_6^{2-}$.

and if one of these is a non-bonding pair it seems probable that only five electron pairs can be comfortably accommodated. Thus sulphur in the SO_2 molecule can be regarded as having a filled valence shell. This is consistent with the known fact that sulphur dioxide does not exhibit any acceptor properties. If an additional electron is added to give the SO_2^- ion it is forced to occupy a secondary valence shell region on the sulphur atom. As this electron is at a greater average distance from the sulphur core than the electrons in the primary valence shell it forms longer bonds than these electrons. Hence the S—S bond in $S_2O_4^{2-}$ is anomalously long. On the other hand, sulphur trioxide is a strong acceptor, and although it has six electron pairs in its valence shell these are all bonding pairs, and as the bonds have considerable polarity this leads to a lower electron density on sulphur than in SO_2, and consequently SO_3 can more readily accommodate an additional electron in its valence shell than SO_2. Thus SO_3^- dimerizes to give an only slightly abnormally long S—S bond in the $S_2O_6^{2-}$ ion. As might be expected, the intermediate $S_2O_5^{2-}$ anion has an intermediate bond length. It is worthy of note that the ClO_2 radical which is isoelectronic with SO_2^- is a stable radical which presumably has a similar electronic structure to SO_2^- and it forms too weak a Cl—Cl bond to dimerize to any measurable extent.

(e) AX_4 Co-ordination: Tetrahedral Geometry

The sulphuryl halides and other SO_2X_2 molecules and their selenium analogues SeO_2X_2 have tetrahedral structures. Bond

Table 7.11 Bond angles for tetrahedral X_2SO_2 molecules

	XSX	OSO	XSO
F_2SO_2	96·1°	124°	
Cl_2SO_2	111·2	119·8	106·5°
$(CH_3)_2SO_2$	112·1	119·4	106·4
$(NH_2)_2SO_2$	115	125	105
$(p\text{BrC}_6H_4)_2SO_2$	100	131	108·7
$C_6H_5SO_2CH_3$	112·0	120·0	107·0

angles are summarized in Table 7.11. The largest angle is always between the two doubly bonded oxygen atoms. Structural data on the sulphate and selenate ions and other molecules also having this tetrahedral structure are given in Fig. 7.26. The small FSF bond

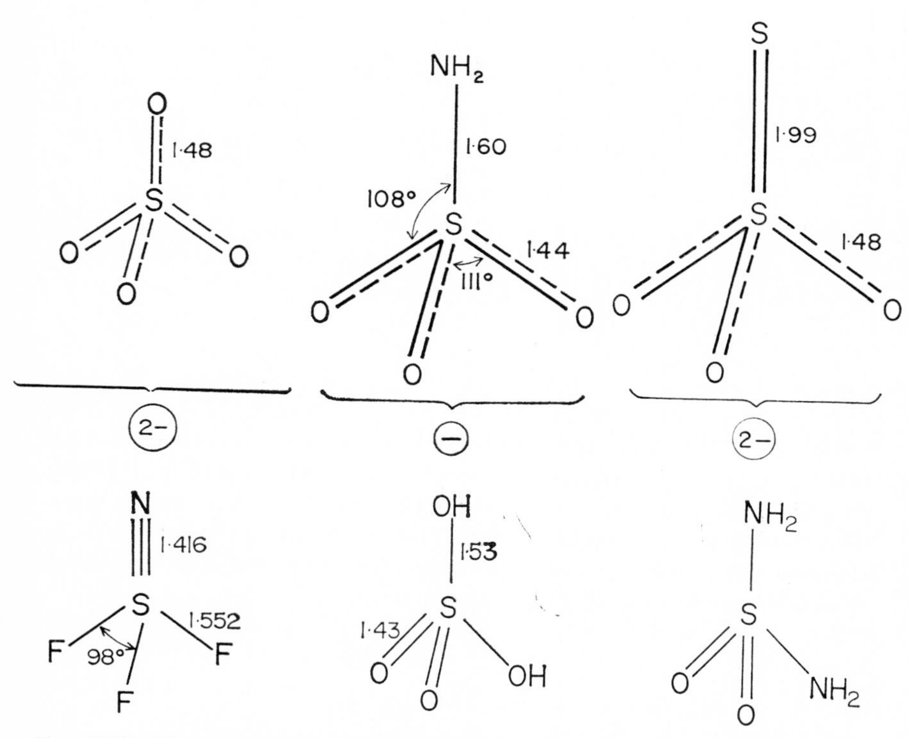

FIG. 7.26 The structures of some tetrahedral AX_4 molecules of sulphur and selenium.

angle in SF_3N is consistent with the strong repulsion exerted by the S≡N triple bond.

(f) AX_4E Co-ordination: Disphenoidal Geometry

The molecule SF_4 has the disphenoid structure expected for a trigonal bipyramidal arrangement of four bonding electron pairs and one non-bonding electron pair in the valence shell (23). The details of the structure of this molecule have been discussed in Chapter 4.

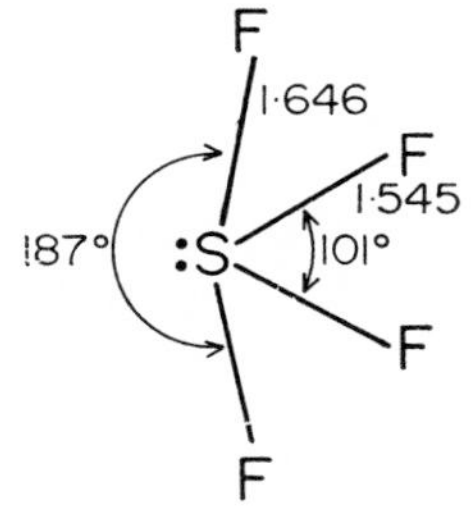

(23)

Structural data on related molecules are given in structures (24) to (27). Other tetrahalides such as SeF_4, $SeCl_4$, and $TeCl_4$ probably have a similar structure but this has not been established with certainty. The only slightly unexpected feature of the structures of (25),

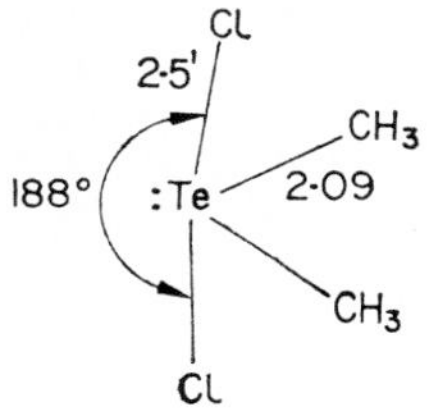

(24) 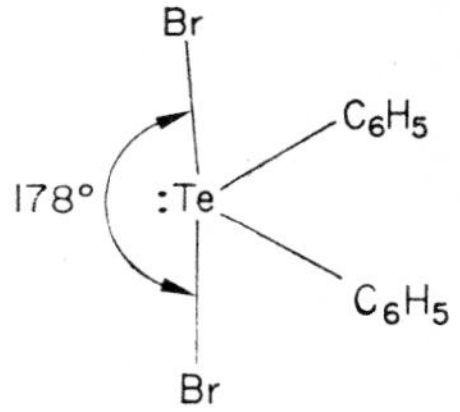 **(25)**

(26), and (27) is that the axial–axial bond angle is less than 180° instead of being slightly greater than 180° as in SF_4 and $(CH_3)_2TeCl_2$. The simplest explanation would be that there is some steric interaction between the bulky phenyl and halogen ligands that forces them slightly apart, thus decreasing the angle between the axial halogen ligands to slightly less than 180°.

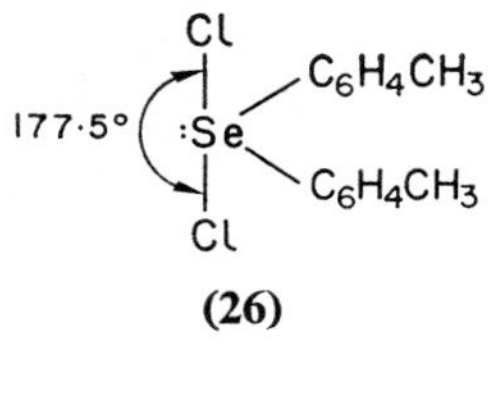

(26)

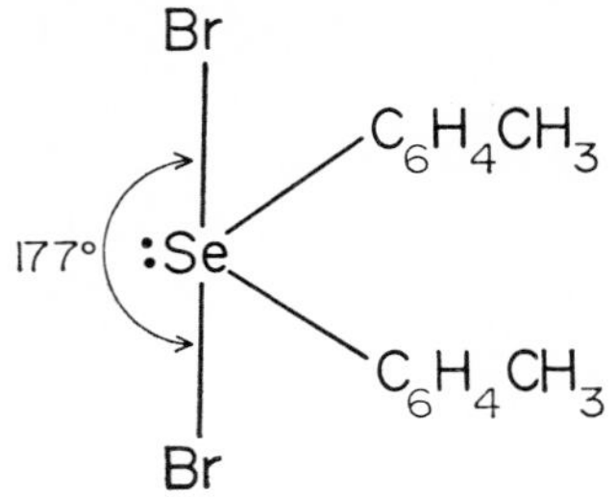

(27)

(g) AX_5 Co-ordination: Trigonal Bipyramidal Geometry

The only molecule of this type whose structure has been determined is SOF_4 (28). As expected, the doubly bonded oxygen occupies an equatorial position, and the large S=O double bond repels the other

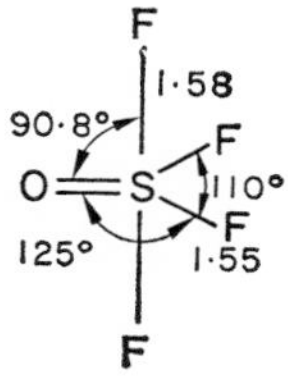

(28)

single-bond pairs so that the angles involving oxygen are larger than the angles between the fluorines, and as usual the axial bonds are longer than the equatorial S—F bonds.

(h) AX_4E_2 Co-ordination: Square Planar Geometry

This stereochemistry is known only in the tellurium (II) complexes with thiourea and tetramethylthiourea, e.g., $Te(SC(NH_2)_2)_2Cl_2$. In these complexes tellurium has a valence shell of six electron pairs with an octahedral arrangement. The two lone-pairs occupy the *trans* positions giving a square planar geometry for the complex.

(i) AX_5E Co-ordination: Square Pyramidal Geometry

This geometry is found for $SeOCl_2.2py$ (29) and for TeF_5^{2-} (30). In TeF_5^{2-} the equatorial bonds are longer than the axial bonds and the axial-equatorial angle is less than 90° as expected (see p. 44). TeF_5^- is isoelectronic with IF_5 and XeF_5^+ and all three molecules have very similar structures. TeF_4 has a fluorine bridged structure with square pyramidal geometry around tellurium (Fig. 7.27).

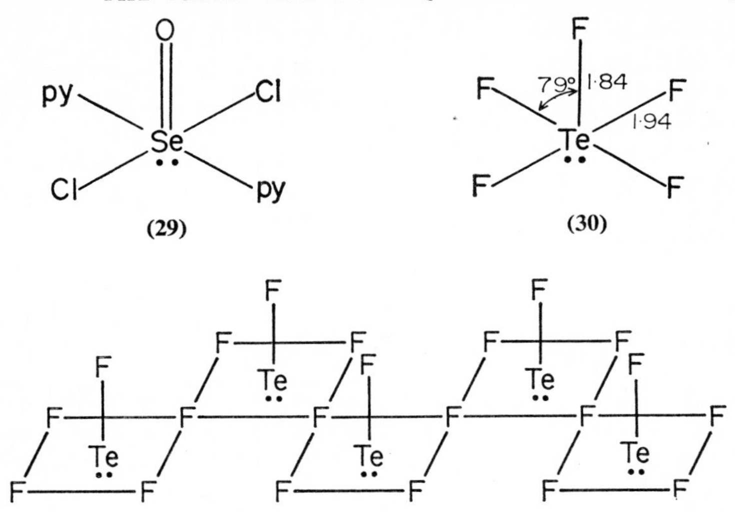

FIG. 7.27 The structure of $(TeF_4)_n$.

(j) AX_6 Co-ordination: Octahedral Geometry

SF_6, SeF_6, TeF_6, and $Te(OH)_6$ are all simple octahedral molecules. S_2F_{10} contains two octahedrally co-ordinated sulphur atoms with an S—S bond. Presumably because of repulsive interaction between the sulphur–fluorine bonds on the two sulphur atoms the fluorine atoms on one sulphur adopt a staggered arrangement with respect to those on the other sulphur. Related molecules with octahedral co-ordination include SF_5Cl, SF_5OF, Te_2F_{10}, F_5SOOSF_5, and $F_5TeOOTeF_5$.

(k) AX_6E Co-ordination: Octahedral Geometry

A number of complex chlorides, bromides, and iodides of selenium and tellurium (IV) are known, e.g., $SeCl_6^{2-}$, $TeBr_6^{2-}$. Despite the fact that their valence shells contain six bonding and one non-bonding electron pair they all appear to have regular octahedral structures. These compounds thus constitute one of the few exceptions to the rules outlined in the earlier chapters of this book. In these compounds the lone-pair appears to have no stereochemical effect, and it must be assumed that it occupies a spherical orbital that lies closer to the nucleus than the orbitals occupied by the bonding electron pairs.

As the central core of an atom increases in size, a lone-pair of electrons has an increasing tendency to spread out around the core in a delocalized spherical orbital in order to decrease its average distance from the nucleus and thus decrease its energy; it does this in preference to remaining in a localized orbital in the valence shell at approximately the same average distance from the nucleus as the other bonding pairs. This tendency will be accentuated in any valence shell that contains a large number of bonding pairs which prevent the lone-pair from occupying very much of the space around the core. In such a case the molecule can apparently reach an overall minimum energy if the lone-pair moves in to occupy a spherical orbital that is then closer to the core than the remaining bonding pairs in the valence shell, and these are forced to move to a somewhat greater distance from the nucleus than they normally would be. This tendency for the lone-pair to move towards the central core has been discussed previously (p. 45), and the situation that we are now discussing is just the limiting case of this general tendency. It should be

Table 7.12 Bond lengths in SeX_6^{2-}, TeX_6^{2-} ions

$TeBr_6^{2-}$	2·62 (2·51) Å	$SeCl_6^{2-}$	2·41 (2·16) Å
$TeCl_6^{2-}$	2·51 (2·38) Å	$SeBr_6^{2-}$	2·54 (2·31) Å

Note: the sum of the covalent radii is given in brackets.

noted that not all AX_6E molecules have this regular octahedral structure, as it appears that $BiBr_6^{3-}$, IF_6^- and XeF_6, for example, definitely have non-octahedral structures. In the case of the fluorides it would seem that the smaller electron pairs of the bonds to fluorine allow sufficient room for the lone-pair in the valence shell.

As the presence of the lone-pair in the inner spherical orbital causes the bonding pairs to be further from the nucleus than they otherwise would be the bond lengths are in all cases abnormally long (Table 7.12). We may note also that when the number of ligands is smaller, as in the SeX_4E and TeX_4E compounds discussed above, the non-bonding pair always remains in the valence shell.

7.6 CHLORINE, BROMINE, AND IODINE

Because of the large number of oxidation states that are known for these elements a large number of different molecular geometries are possible; these are summarized in Table 7.13.

(a) AX_2E_2 Co-ordination: Angular Geometry

The cations ICl_2^+, BrF_2^+, ClF_2^+, Cl_2F^+, and Cl_3^+ are known to be angular (Table 7.14). Other related cations, e.g. $I(pyridine)_2^+$, I_3^+, and Br_3^+ presumably have similar bent structures. The bond angles

Table 7.13 Observed molecular geometries for chlorine, bromine, and iodine

Number of lone-pairs plus bonds	Arrangement	Lone-pairs	Bonds		Geometry	Example
4	Tetrahedron	2	2	AX_2E_2	Angular	ICl_2^+
		1	3	AX_3E	Pyramidal	ClO_3^-
		0	4	AX_4	Tetrahedral	ClO_3F
5	Trigonal bipyramid	3	2	AX_2E_3	Linear	ICl_2^-
		2	3	AX_3E_2	T-shaped	ClF_3
		1	4	AX_4E	Disphenoidal	$IO_2F_2^-$
		0	5	AX_5	Trigonal bipyramidal	
6	Octahedron	2	4	AX_4E_2	Square planar	ICl_4^-
		1	5	AX_5E	Square pyramidal	BrF_5
		0	6	AX_6	Octahedral	IOF_5
7	Pentagonal bipyramid	1	6	AX_6E	Distorted octahedral	IF_6^-
		0	7	AX_7	Pentagonal bipyramidal	IF_7

are, as expected, rather less than the tetrahedral angle. There is strong bridge bonding by halogens in these compounds, so that the angular molecule becomes part of an approximately octahedral arrangement of electron pairs around the central atom, in which it is

Table 7.14 Bond angles in $AX_2E_2^+$ cations

$ICl_2^+SbCl_6^-$	92·5°
$ICl_2^+AlCl_4^-$	96·7
$ClF_2^+AsF_6^-$	103·2
$ClF_2^+SbF_6^-$	95·9
$BrF_2^+SbF_6^-$	93·5
$Cl_3^+AsF_6^-$	~100°

surrounded by four ligands in an approximately square planar arrangement and two *trans* lone-pairs as shown in Fig. 7.28. Hence the bond angle has a tendency to approach the 90° angle of the octahedron.

The ClO_2^- ion also has this angular geometry with a bond angle of 110·5° and a bond length of 1·56 Å. The bond angle is quite close to the tetrahedral angle that would be expected if the bonds, which presumably have a bond order of about 1·5, and the lone-pairs repel each other to approximately the same extent. When one electron is removed to form the stable ClO_2 radical, the bond angle increases to 117·4° and the bond length decreases to 1·47 Å. The decrease in bond length can be attributed to a greater Cl—O bond order, and the

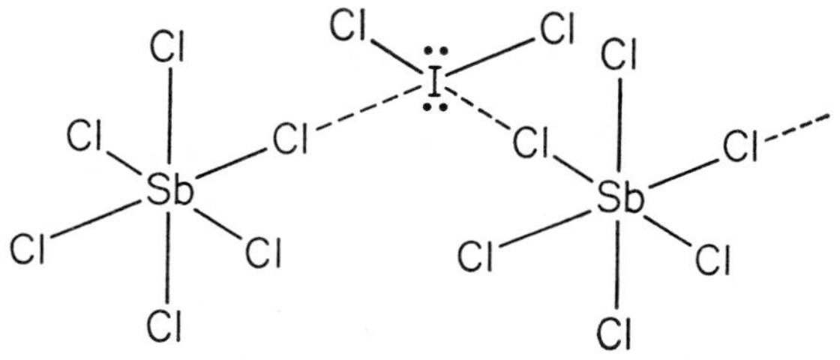

FIG. 7.28 The structure of $ICl_2.SbCl_6$.

increased bond angle is due to this greater bond order and to the fact that there are only three non-bonding electrons instead of four. In fact it seems reasonable to propose that the bonds are essentially full double bonds and that the single unpaired electron occupies an orbital in the secondary valence shell as in the isoelectronic SO_2^- ion. This electron is then at a greater distance from the nucleus than the electrons in the primary valence shell and apparently cannot form a strong enough bond with another ClO_2 molecule to lead to any appreciable dimerization.

1·57 Cl 1/2 O 110·5° O   Cl O O

(b) AX_3E Co-ordination: Pyramidal Geometry

This type of co-ordination is found in the chlorate, bromate, and iodate ions and in iodic acid, all of which have pyramidal structures (31–34). The bond angle in the iodate ion is approximately 97° and in chlorate it is 106·7°. In BrO_3^- the angle has been given as 112°, but the error in the value is uncertain. The angle would be expected to increase in the series $IO_3^- < BrO_3^- < ClO_3^-$. Chloryl fluoride would also be expected to have this pyramidal structure (35).

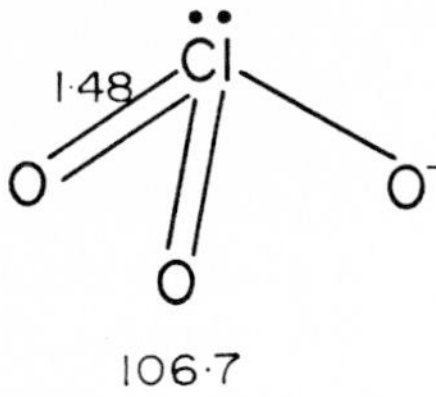

1·48 Cl O O O⁻
106·7

(31)

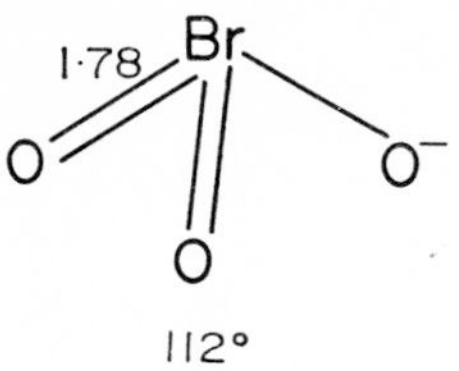

1·78 Br O O O⁻
112°

(32)

1·82 I O O O⁻
97°

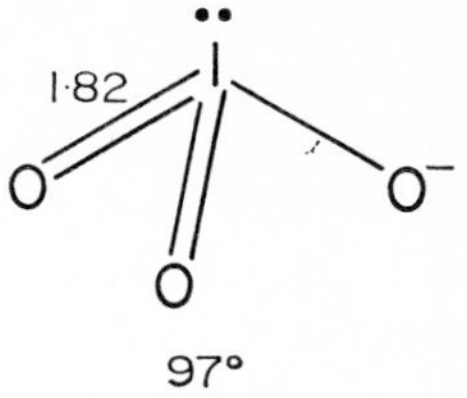

(33)

(c) AX_4 Co-ordination: Tetrahedral Geometry

The perchlorate, perbromate, and periodate ions ClO_4^-, BrO_4^-, and IO_4^- are tetrahedral. The decrease in the bond length in the series ClO_2^- 1·56 Å, ClO_3^- 1·46, and ClO_4^- 1·42 is consistent with

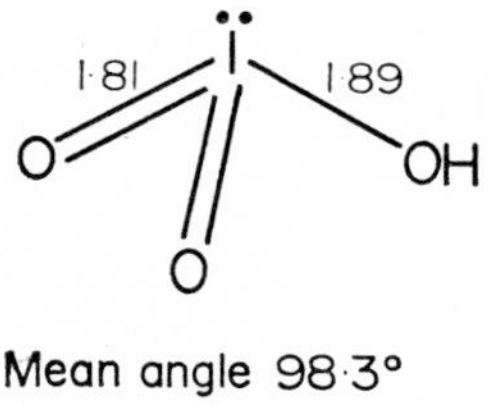

1·81 I 1·89 O O OH
Mean angle 98·3°

(34)

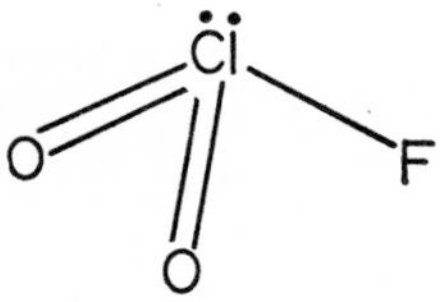

Cl O O F

(35)

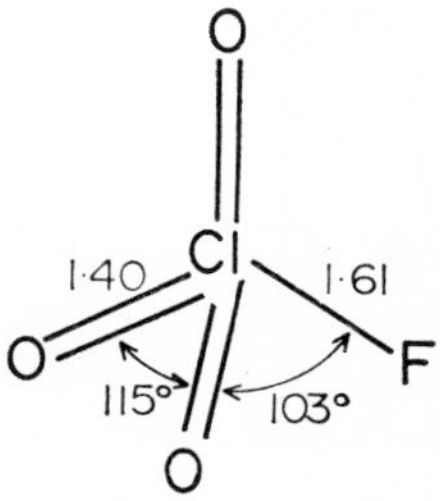

O Cl 1·40 1·61 O O F 115° 103°

(36)

the increasing bond order in this series and with the increasing number of electronegative ligands. Other molecules with tetrahedral structures of this type include perchloryl fluoride $FClO_3$ (36), perchloric acid $HOClO_3$ (37), and dichlorine heptoxide Cl_2O_7 (38).

(d) AX_2E_3 Co-ordination: Linear Geometry

In this case there are two axial bonding electron pairs and three equatorial non-bonding pairs. This shape is found for a number of trihalide ions, e.g., I_3^-, ICl_2^-, IBrCl, and IBr_2^- (39–41).

With large cations the I_3^- ion is linear and symmetrical, but with smaller cations, e.g., NH_4^+ and Cs^+, it is somewhat distorted. The reason for this distortion is not certain, although it may arise simply

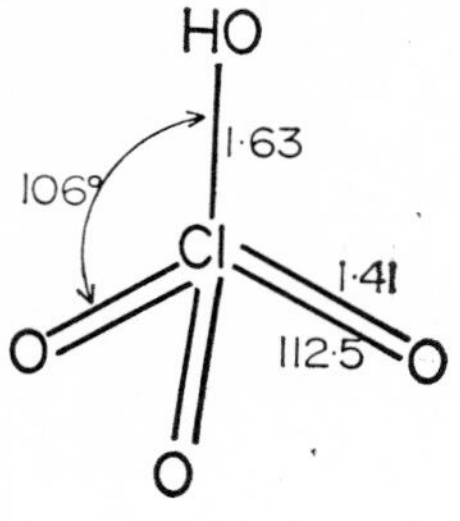

HO
1·63
106°
Cl
1·41
112·5
O O O

(37)

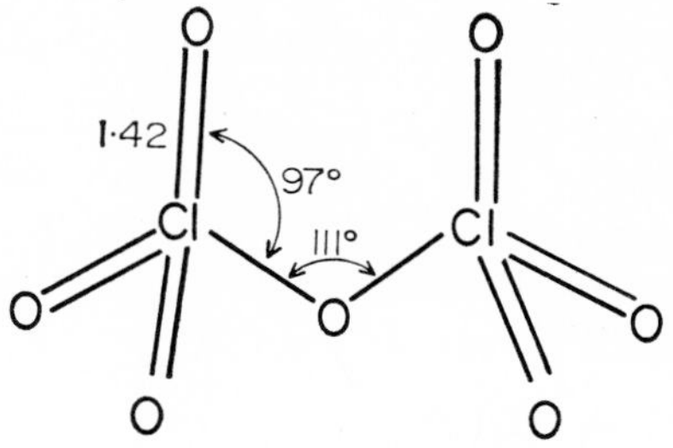

O O
1·42
97°
Cl 111° Cl
O O O O O

(38)

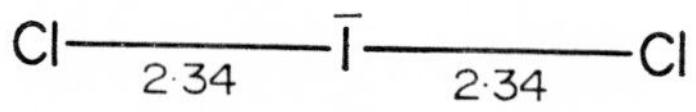

Cl—2·34—Ī—2·34—Cl

(39)

Br—2·50—Ī—2·38—Cl

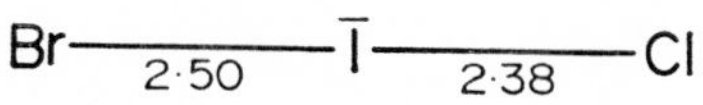

(40)

from crystal packing considerations. A similar distortion occurs in Cs_2I_8 in which two I_3^- ions are linked together by an iodine molecule (42).

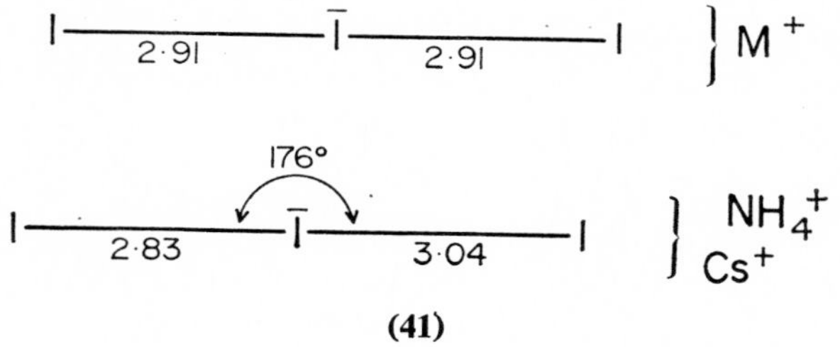

I—2·91—Ī—2·91—I } M^+

176°

I—2·83—Ī—3·04—I } NH_4^+ Cs^+

(41)

I
I 80° I 2·83
177° I
2·86
I I I
3·42 3·00
I

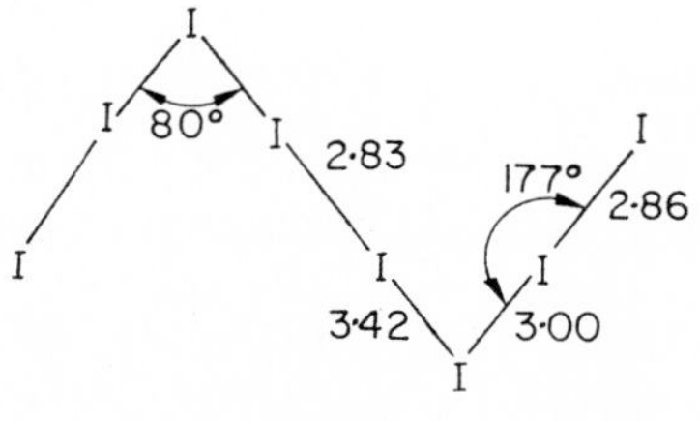

(42)

(e) AX_3E_2 Co-ordination: T-shaped Geometry

This shape is found for the molecules ClF_3, BrF_3, and $C_6H_5ICl_2$ (43–45). The valence shell contains five electron pairs with two non-bonding pairs in the equatorial positions of a trigonal bipyramid. The bond lengths and bond angles are consistent with the greater repulsion exerted by the lone-pairs as explained in Chapter 4.

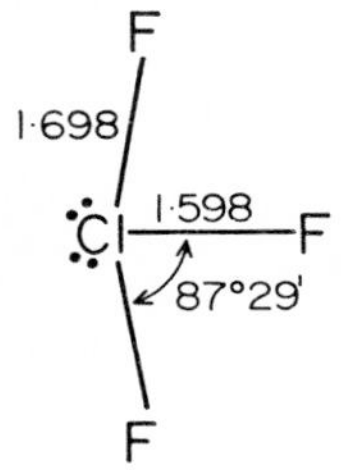

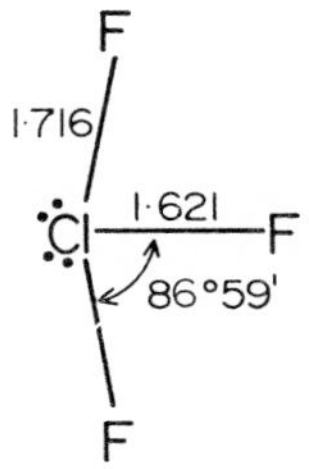

gas phase (microwave) | solid state (x-ray)

(43) (44) 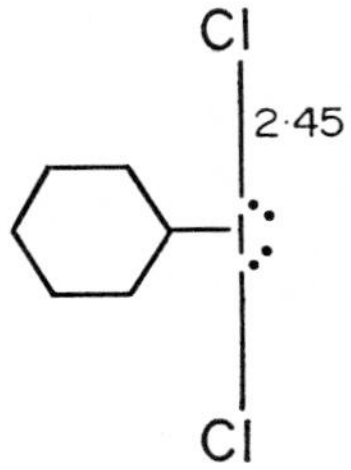 (45)

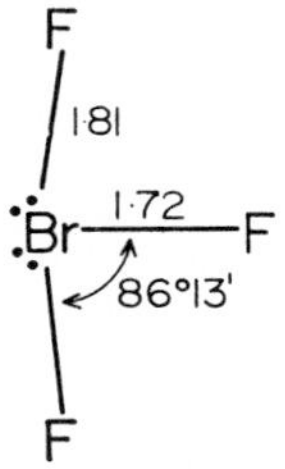

(f) AX_4E Co-ordination: Disphenoidal Geometry

This geometry which is just like that of SF_4 is found in $IO_2F_2^-$ (46) and IF_4^+ (47). The latter ion is present in the compound $IF_5 . SbF_5$ which has the ionic $IF_4^+SbF_6^-$ structure.

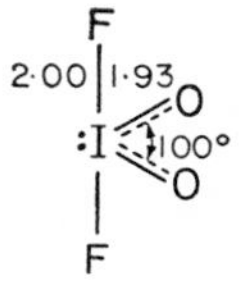

(46) 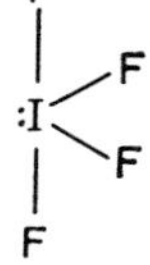 (47)

(g) AX_5 Co-ordination: Trigonal Bipyramid Geometry

IO_2F_3 has been prepared but its structure has not yet been determined. It may be predicted to be trigonal bipyramidal with the two oxygen atoms in equatorial positions.

(h) AX_4E_2 Co-ordination: Square Planar Geometry

The ICl_4^- and BrF_4^- ions and the I_2Cl_6 molecule have this geometry (48–50). In I_2Cl_6 the bonds to the bridging chlorines are

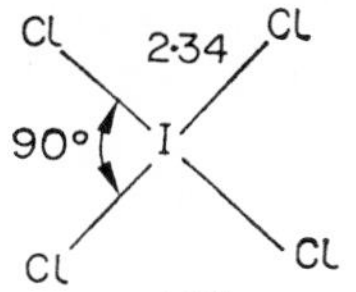

(48)

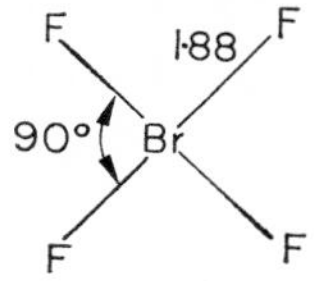

(49)

longer than those to the terminal chlorines. The bridging chlorines carry a formal charge of +1 and are therefore rather more electronegative than the terminal chlorines, hence they attract the bridge bond electron pairs more strongly, and these will occupy less of the

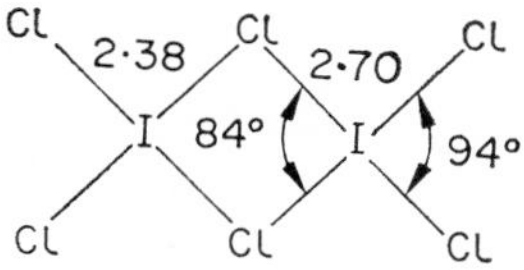

(50)

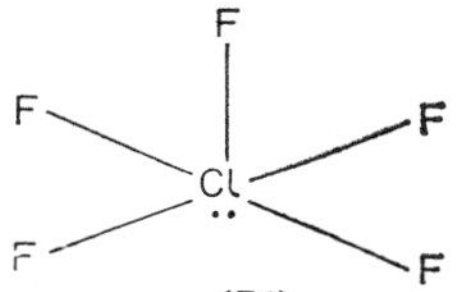

(51)

surface of the iodine than the terminal bond electron pairs; hence the angle between the terminal bonds is greater than that between the bridge bonds at iodine.

(i) AX_5E Co-ordination: Square Pyramidal Geometry

ClF_5, BrF_5, and IF_5 have this shape (51–53). The structure of ClF_5 has been established from spectroscopic data and the

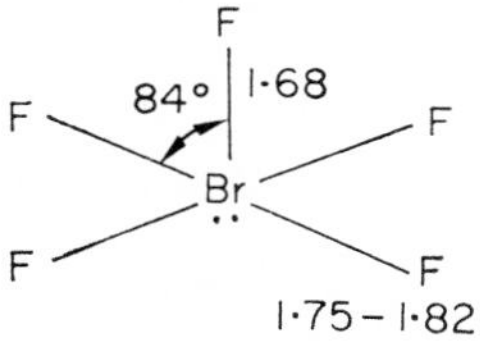

(52)

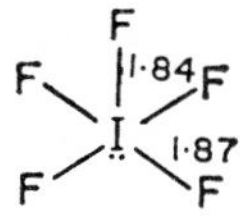

(53)

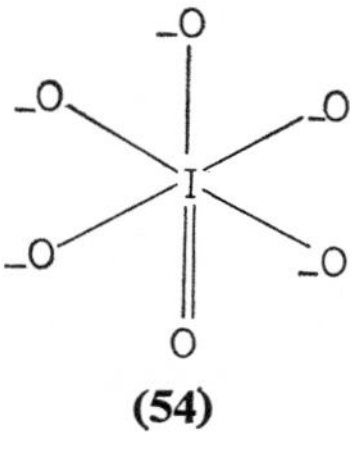

(54)

molecular parameters are not known. For BrF_5 and IF_5 the expected deviation from a regular square pyramid structure due to the lone-pair is observed.

(j) AX_6 Co-ordination: Octahedral Geometry

The periodate ion IO_6^{5-}, iodine oxopentafluoride IOF_5 and and IF_6^+ have the expected octahedral structure (54–56).

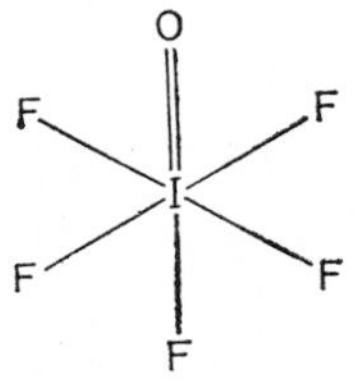

(55)

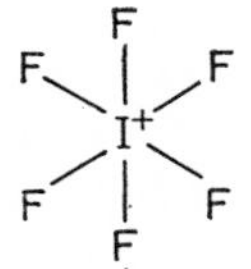

(56)

(k) AX_6E Co-ordination: Distorted Octahedral Geometry

The ions IF_6^- and BrF_6^- are of this type but their structures are not known with certainty. Infra-red and Raman spectra of IF_6^- suggest that it is not octahedral. The arrangement of seven electron pairs cannot be predicted with complete certainty. The most likely arrangements are discussed in Chapter 5. In view of the existence of IF_7 it is reasonable to suppose that there is room for the lone-pair to occupy the valence shell and exert a stereochemical effect, so it is expected that IF_6^- will not have an octahedral structure. The most likely structure is that which minimizes the number of lone-pair–bond-pair interactions and this would appear to be the 1:3:3 structure with the lone-pair in the middle of an octahedral face which will of course be distorted.

(l) AX_7 Co-ordination: Pentagonal Bipyramidal Geometry

The only molecule of this class that is known is IF_7. There has been considerable controversy over the structure of this molecule, but the consensus of opinion appears to be that the structure is an approximate pentagonal bipyramid in which there is some buckling of the equatorial ring of fluorines. This is probably an example of a stereochemically non-rigid molecule. The existence of at least two other stereochemical forms which do not have a very much higher energy than the pentagonal bipyramid allows the pentagonal bipyramid to undergo an internal rearrangement to an equivalent pentagonal bipyramidal form with a rather low activation energy, in a manner analogous to the previously described pseudo-rotation for trigonal bipyramid molecules (p. 69).

7.7 NOBLE GAS COMPOUNDS

Since the discovery of the first xenon compound by N. Bartlett in 1962 there has been a rapid and extensive development of this area of chemistry and a number of compounds have been prepared and their structures determined. The unexpected stability of these noble gas compounds led many chemists to assume that the nature of the bonding in these compounds was rather unusual. However, this appears not to be the case and apparently they all have ordinary covalent bonds and their structures are completely in accord with the principles discussed in this book. Indeed, these compounds demonstrate very clearly the ease with which the structures of molecules may be predicted.

Table 7.14 The molecular geometries of xenon compounds

Number of lone-pairs and bonds	Arrangement	Number of lone-pairs	Geometry		Examples
4	Tetrahedral	2	AX_2E_2	angular	XeO_2*
		1	AX_3E	pyramidal	XeO_3
		0	AX_4	tetrahedral	XeO_4
5	Trigonal bipyramidal	3	AX_2E_3	linear	XeF_2 $FXeOSO_2F$ KrF_2
		2	AX_3E_2	T-shaped	$XeOF_2$*
		1	AX_4E	disphenoidal	XeO_2F_2
		0	AX_5	trigonal bipyramidal	XeO_3F_2*
6	Octahedral	2	AX_4E_2	square planar	XeF_4
		1	AX_5E	square pyramidal	$XeOF_4$ XeF_5^+ $(XeO_3F^-)_n$
		0	AX_6	octahedral	XeO_6^{4-} XeO_2F_4*
7	Monocapped octahedron	1		distorted octahedral	XeF_6

* Predicted shapes.

The molecular parameters in those cases where the structure has been determined are given in Fig. 7.29. The tetrahedral nature of XeO_4 has been established from its vibrational spectrum. Xenon trioxide has the expected pyramidal structure, and we note that although the XeO bonds are best formulated as double bonds, the angle between them is quite small. This is however consistent with

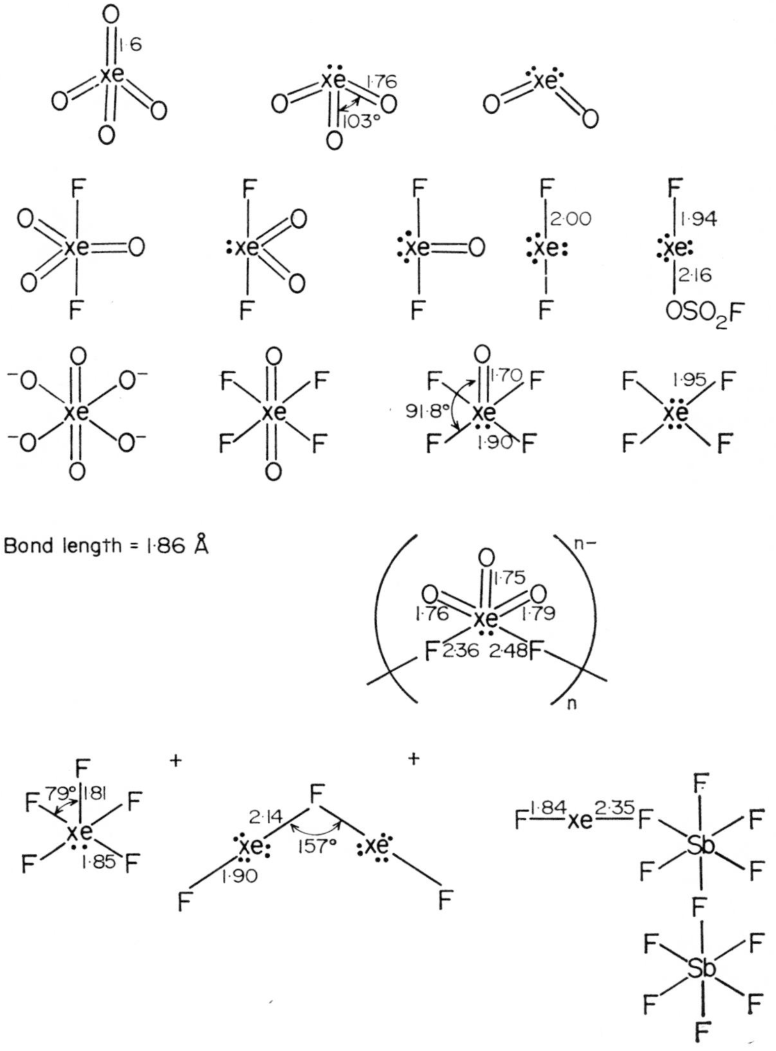

Bond length = 1·86 Å

FIG. 7.29 The structures of some xenon compounds.

the general rule that as the size of the central atom increases and hence the average distance between the bonds increases, repulsions between the bonds decrease. Thus the lone-pair is able to expand to take up more room and the bond angles decrease. Of the molecules containing a total of five bonds and lone-pairs in the valence shell of the central atom XeF_2, $FXeOSO_2F$, and KrF_2 have the expected linear structure with three lone-pairs occupying the equatorial positions. Of the related oxyfluorides only XeO_2F_2 has been established with certainty, and spectroscopic data indicate the structure shown. It may be predicted that the oxyfluorides XeO_3F_2 and $XeOF_2$ would have closely related structures with the oxygen atoms and lone-pairs occupying the equatorial positions of a trigonal bipyramid arrangement (Fig. 7.29). We note that the Xe—F bonds in XeF_2 are axial bonds that are interacting with three equatorial lone-pairs. They may therefore be expected to be somewhat abnormally long and this is at least part of the reason why they are longer than the bonds in XeF_4.

The perxenate ion, XeO_6^{4-} ion, has been established to have the expected octahedral structure and $XeOF_4$ to have the expected square pyramid structure. In contrast to XeF_5^+ where the bond angle between axial and equatorial bonds is only 79° the O═Xe—F bond angle in $XeOF_4$ is slightly greater than 90°, which indicates that the repulsive effect of the Xe═O double bond is slightly greater than that of the lone-pair in this compound. The anion XeO_3F^- is polymeric and consists of infinite fluorine bridged chains giving square pyramidal AX_5E co-ordination around xenon with one of the oxygens occupying the axial position (Fig. 7.29). The molecule XeO_2F_4 is not known, but if it is prepared it certainly would be expected to have an octahedral structure with *trans* oxygens. The square planar structure of the AX_4E_2 molecule XeF_4 is known very accurately. The Xe—F bond length is appreciably shorter than in XeF_2, and in addition to the reason given above we may note that this is in accordance with the general rule that bond lengths decrease with increasing effective electronegativity of the central atom.

There has been much discussion on the shape of XeF_6, but it now seems to be clearly established that it is not a regular octahedral molecule but that the arrangement of the fluorines is distorted from the regular octahedral arrangement in such a way as to indicate that the non-bonding electron pair occupies a position in the centre of one face of an octahedron so that the fluorines at the corners of this face are forced apart somewhat (Fig. 7.30).

Four cationic xenon fluorine compounds have also had their

structures established. XeF_5^+ in $XeF_5^+ . PtF_6^-$ is an AX_5E molecule with the expected square pyramid structure. In accordance with expectations, the bond angle between the basal bonds and the axial bond is less than the ideal angle of 90°, and the basal bonds are

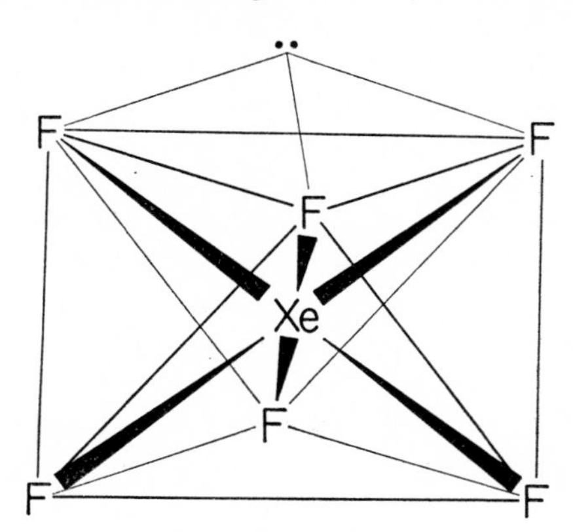

FIG. 7.30 The structure of XeF_6

somewhat longer than the axial bond as a consequence of the repulsion exerted by the lone-pair in the sixth octahedral position. The ion $Xe_2F_3^+$ in $Xe_2F_3 . AsF_6$ has the expected linear geometry around xenon as in XeF_2 and an angular geometry at the central fluorine. The compound $XeF_2 . 2SbF_5$ has been shown to contain XeF^+ and $Sb_2F_{11}^-$ but the XeF^+ ion is rather strongly bonded to the $Sb_2F_{11}^-$ by a fluorine bridge. The short length of the bridge bond indicates considerable covalent character so that it cannot be completely satisfactorily described as an ionic compound. The two bonds at the xenon are at 180° as expected. The compound $XeF_4 . 2SbF_5$ has been shown to have the ionic structure $XeF_3^+ . Sb_2F_{11}^-$. The XeF_3^+ ion has the expected T-shaped structure.

REFERENCES AND SUGGESTIONS FOR FURTHER READING

Tables of Interatomic Distances and Configuration in Molecules and Ions, Special Publication No. 11, The Chemical Society, London (1958).

Supplement, Special Publication No. 18, The Chemical Society, London (1965).

F. A. COTTON and G. WILKINSON, *Advanced Inorganic Chemistry*, 2nd Ed., Interscience, 1966.

R. J. GILLESPIE, *J. Amer. Chem. Soc.*, **82,** 5978 (1960).

J. H. HOLLOWAY, *Noble Gas Chemistry*, Methuen, 1968.

L. PAULING, *Nature of the Chemical Bond*, 3rd Ed., Cornell University Press, 1960.

A. F. WELLS, *Structural Inorganic Chemistry*, 3rd Ed., Oxford University Press, 1962.

8

Transition Elements

8.1 NON-BONDING *d* SHELLS

The electron shells of an atom can be further subdivided into sub-shells, each sub-shell consisting of a number of equivalent orbitals each of which can contain two electrons of opposite spin. These sub-shells can accommodate 2, 6, 10, 14 . . . electrons and thus consist of 1, 3, 5, 7 . . . orbitals respectively which are known as *s*, *p*, *d*, *f* . . . orbitals. Thus the K shells consists of a single 1*s* orbital, the L shell of a 2*s* orbital and three 2*p* orbitals, the M shell of a 3*s* orbital, three 3*p* orbitals and five 3*d* orbitals, etc. In a polyelectronic atom these sub-shells do not all have the same energy but increase in energy in the series $s < p < d < f$ . . . This gives rise to the possi-

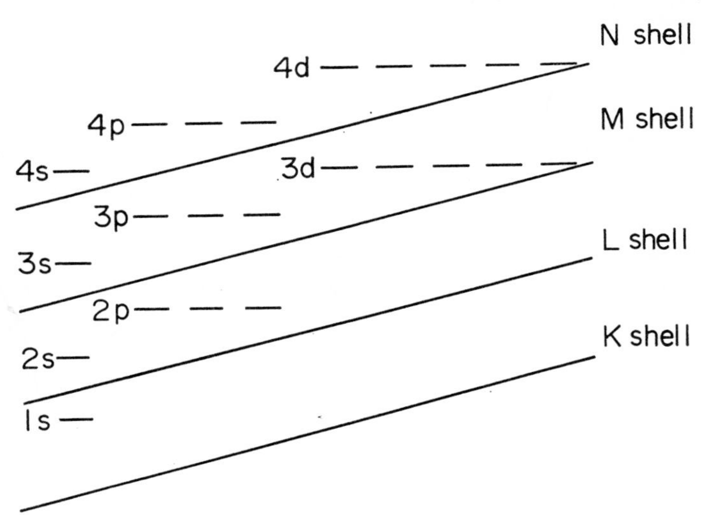

FIG. 8.1 Approximate energy level diagram for polyelectronic atoms.

bility that the sub-shells of different main shells may overlap in energy as shown in the very approximate energy level diagram given in Fig. 8.1. This diagram is only approximate as the relative energies of the different shells and sub-shells vary from atom to atom as the nuclear charge and the interelectronic repulsions change. The first important consequence of this overlap of energy levels is that the 3*d* orbitals are of higher energy than the 4*s* orbital at least for the elements from potassium to nickel so that the 4*s* orbital is filled with one electron in potassium and with two electrons in calcium and then the

Table 8.1 Electronic configurations of the elements 19–36

Electron shell	K	L		M			N	
Sub shell element	1*s*	2*s*	2*p*	3*s*	3*p*	3*d*	4*s*	4*p*
K	2	2	6	2	6		1	
Ca	2	2	6	2	6		2	
Sc	2	2	6	2	6	1	2	
Ti	2	2	6	2	6	2	2	
V	2	2	6	2	6	3	2	
Cr	2	2	6	2	6	5	1	
Mn	2	2	6	2	6	5	2	
Fe	2	2	6	2	6	6	2	
Co	2	2	6	2	6	7	2	
Ni	2	2	6	2	6	8	2	
Cu	2	2	6	2	6	10	1	
Zn	2	2	6	2	6	10	2	
Ca	2	2	6	2	6	10	2	1
Ge	2	2	6	2	6	10	2	2
As	2	2	6	2	6	10	2	3
Se	2	2	6	2	6	10	2	4
Br	2	2	6	2	6	10	2	5
Kr	2	2	6	2	6	10	2	6

3*d* orbitals begin to fill with scandium and are finally completely filled at zinc (Table 8.1). These elements thus have from one to twelve electrons outside the argon, krypton or xenon core, not all of which are necessarily used in bonding. Frequently some of the *d* orbital electrons remain unused as is evidenced by the frequent occurrence of paramagnetic compounds of these elements resulting from the presence of unpaired electrons in the *d* orbitals. Thus a problem arises as to what constitutes the valence shell for these elements. For potassium and calcium the 4*s* electrons clearly lie outside the spherical argon core consisting of the completed 3*s* and 3*p* sub-shells, and for the elements scandium and titanium, which very easily use not only the 4*s* electrons but also their 3*d* electrons in compound formation, it is clear that the 3*d* and 4*s* electrons can be regarded as constituting the valence shell outside the completed argon core. However, for zinc the 3*d* shell is complete with ten electrons and thus the M shell is also complete. Hence these 3*d* electrons are never used in compound formation and it is clear that they no longer form part of the valence shell. It is reasonable to assume that any 3*d* electrons not used in compound formation occupy an inner shell inside the valence shell which can interact to a greater or lesser extent with the valence shell, depending on the number of electrons in this inner *d* shell, the nuclear charge (i.e., the atomic number) and the number of electrons in the outer valence shell.

Table 8.2 Geometry of the d^0, d^5, and d^{10} transition elements

Co-ordination number	Shape	Number of *d* electrons	Example
2	Linear	10	$Ag(NH_3)_2^+$
3	Equilateral triangle	5	$Fe\{N(SiH_3)_2\}_3$
		10	$[Cu(CN)_2^-]_n$
4	Tetrahedron	0	$TiCl_4$
		5	$FeCl_4^-$
		10	$ZnCl_4^{2-}$
5	Trigonal bipyramid	0	$NbCl_5$
		5	$[Mn(Sal—Me)_2]_2$*
		10	$Zn(acac)_2H_2O$†
6	Octahedron	0	WF_6
		5	FeF_6^{3-}
		10	$Zn(NH_3)_6^{2+}$

* Sal—Me = N-methylsalicylaldiminate

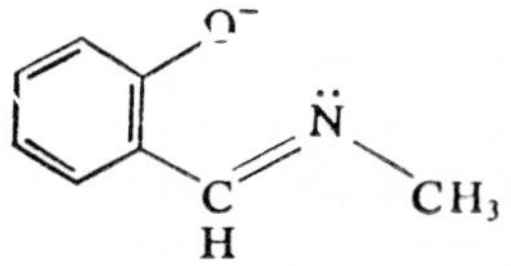

O⁻ N̈ CH₃ C H

† acac = acetylacetonate

O⁻ O C C C C CH₃ H

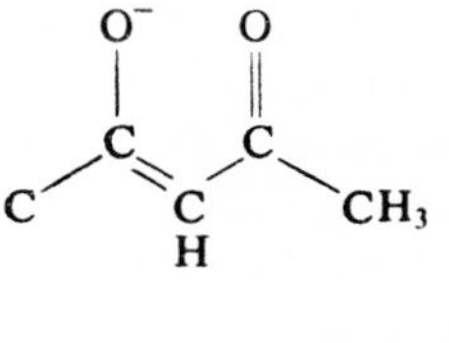

l is symmetrical, i.e., is empty (d^0), is half-filled with ingly occupied (d^5) or is complete (d^{10}) the d shell effect on the arrangement of surrounding valence shell . When the number of d electrons is small, typically one nteraction of the d shell with the bonding electron pairs weak and any distortions of the arrangement of the tron pairs appears to be negligible or is at least too small ed. In cases where the d shell is extensively filled, typically d^9, it generally interacts rather strongly with the bonding ırs and has an effect on their arrangement. This effect can d by assuming that the d shell, which will be non-spherical ses, has an approximately ellipsoidal shape which could be ate or oblate. Thus the stereochemistry of the compounds ınsition elements is typically that of the non-transition elem... n their compounds in which there are no unshared electrons in the valence shell, i.e., AX_2-linear, AX_3-triangular, AX_4-tetrahedral, AX_5-trigonal bipyramidal, AX_6-octahedral, etc., but which are sometimes somewhat distorted by the presence of an underlying non-spherical d shell. As a consequence, the transition elements show rather less variety in the geometry of their molecules than the main group elements, and the tetrahedron and the octahedron are the predominant shapes for transition metal compounds. Table 8.2 summarizes the various geometries observed for transition metal compounds which have an underlying spherical d shell and in which therefore there is no distortion of the expected arrangement of a given number of electron pairs around the central atom.

8.2 TWO CO-ORDINATION: LINEAR AX_2 GEOMETRY

This geometry is found only for the elements copper, silver, and gold in the +1 oxidation state, zinc and cadmium in the +2 oxidation state, and mercury in both the +1 and +2 oxidation states. As these are all d^{10} states no distortions due to a non-spherical d shell are expected or are found.

Two co-linear bonds are particularly common in the compounds of Ag(I), Au(I), Hg(I), and Hg(II). This can be attributed to the rather large size and small charge of the inner cores of these elements which leads to a rather weak electric field at the surface and thus tends to limit the number of electron pairs that they can hold in their valence shells to a small number. The lighter elements in these groups, i.e., copper, zinc, and cadmium have smaller atomic cores, and therefore stronger electric fields, and are generally able to hold more electron pairs in their valence shells and thus they achieve higher co-ordination numbers. Some typical examples of linear two-co-ordinated geometry observed for these elements are given in Table 8.3 and

Table 8.3 AX_2 linear geometry

Hg(II)	ClHgCl, BrHgBr, ClHgSCN, CH_3HgCl, CH_3HgBr, CH_3HgCH_3, NCHgCN (171°), HgO (Fig. 8.3), $HgNH_2Br$ (Fig. 8.3)
Hg(I)	ClHgHgCl, BrHgHgBr, IHgHgI
Ag(I)	$[NCAgCN]^-$, $[H_3NAgNH_3]^+$, $[ClAgCl]^-$, AgCN (Fig. 8.3)
Au(I)	$[NCAuCN]^-$, $[ClAuCl]^-$
Zn(II)	CH_3ZnCH_3, $(CO)_4CoZnCo(CO)_4$

All bond angles are 180° except where noted.

Figs. 8.2 and 8.3. In the very few cases where the bond angle at the metal atom is not 180° this appears to be due to weak co-ordination to other atoms at larger distances, giving what might be regarded as a very distorted tetrahedral arrangement of four bonds, e.g.,

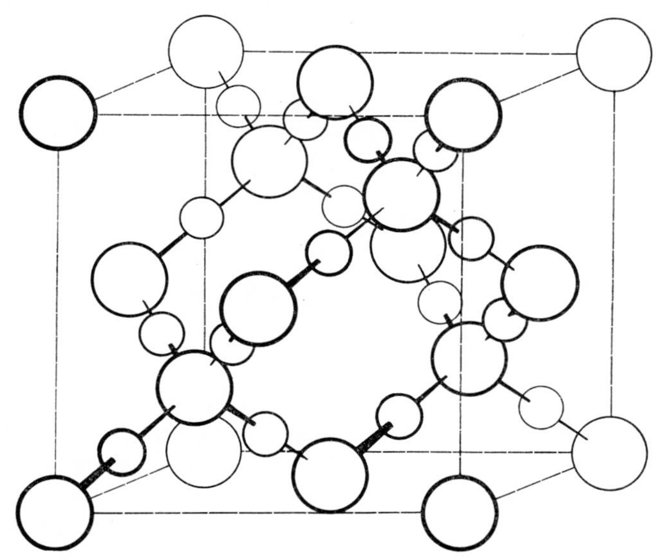

FIG. 8.2 The structure Ag_2O: ○ silver, ◯ oxygen.

$Hg(CN)_2$. In silver oxide each silver atom has linear two-co-ordination and each oxygen atom has tetrahedral four-co-ordination (Fig. 8.2). In HgO, $HgNH_2Br$, and AgCN there are infinite chains with the structures shown in Fig. 8.3.

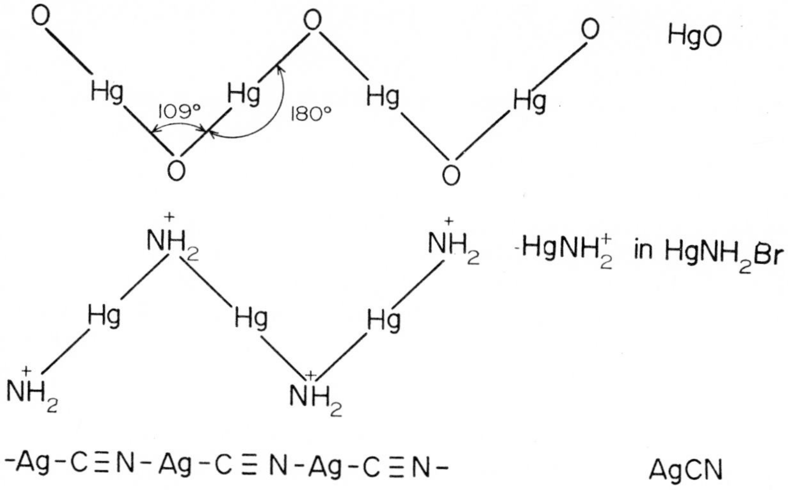

O Hg O Hg O Hg O Hg O HgO

109° 180°

$\overset{+}{N}H_2$ Hg $\overset{+}{N}H_2$ Hg $\overset{+}{N}H_2$ Hg $\overset{+}{N}H_2$ $-HgNH_2^+$ in $HgNH_2Br$

$-Ag-C\equiv N-Ag-C\equiv N-Ag-C\equiv N-$ AgCN

FIG. 8.3 The structures of some polymeric molecules containing linear 2-co-ordinated atoms.

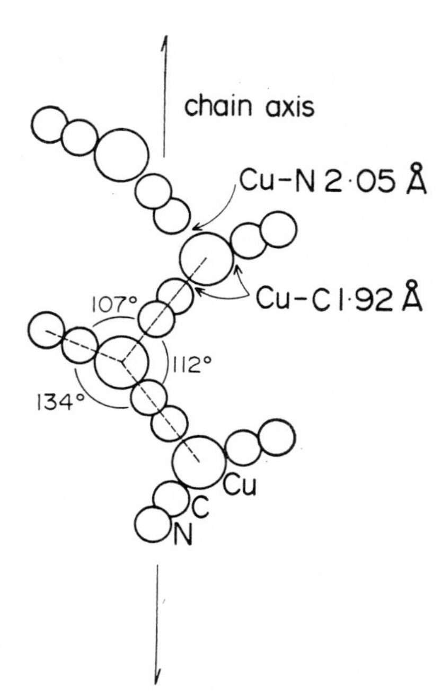

FIG. 8.4 The structure of copper(II) cyanide $Cu(CN)_2$. The molecule is an infinite chain polymer.

8.3 THREE CO-ORDINATION: TRIGONAL PLANAR AX_3 GEOMETRY

This is an uncommon type of geometry and there appear to be only two clearly established examples. Copper(II) cyanide has a helical chain structure with the geometry shown in Fig. 8.4. The distortions of the expected bond angles of 120° around copper are consistent with the fact that the Cu—N bond is longer and weaker than the Cu—C bond. This trigonal planar geometry would be unaffected by the d^9 non-bonding shell as a prolate ellipsoidal d-shell has a circular cross-section in the molecular plane. The molecule $Fe[N(SiH_3)_2]_3$ which has a d^5 non-bonding shell has the expected trigonal planar structure.

8.4 FOUR CO-ORDINATION: AX_4 TETRAHEDRAL GEOMETRY

This is a common geometry for the transition elements in their compounds, and is found for the symmetrical d^0, d^5, and d^{10} shells, and also for some non-spherical d shells.

Some examples of tetrahedral molecules are given in Table 8.4. In general tetrahedral four-co-ordination is commonest for oxo molecules and complexes and for some of the transition elements in

Table 8.4 Tetrahedral AX_4 geometry

d^0	$TiCl_4$, $ZrCl_4$, $VOCl_3$, CrO_2F_2, CrO_2Cl_2, CrO_3Cl^- CrO_4^{2-}, OsO_4, OsO_3N^-, MnO_4^-, MoO_2Cl_2, MnO_3F
d^1	VCl_4, RuO_4^-
d^2	VCl_4^-, RuO_4^{2-}
d^5	$MnCl_4^-$, $FeCl_4^-$
d^6	$FeCl_4^{2-}$
d^7	$CoCl_4^{2-}$, $Co(NH_3)_4^{2+}$
d^8	$NiCl_4^{2-}$, $NiCl_2.(PPh_3)_2$
d^{10}	$Cu(CN)_4^{3-}$, $Cu\left(SC\begin{smallmatrix}NH_2\\CH_3\end{smallmatrix}\right)_4^+Cl^-$ $ZnCl_4^{2-}$, $Zn(CN)_4^{2-}$, $CdBr_4^{2-}$, $Hg(SCN)_4^{2-}$

their lower oxidation states where the small charge of the atomic core tends to limit the number of electron pairs that can be held in the valence shell. The cuprous halides and silver iodide have the zinc blende structure with four-co-ordination around both copper on

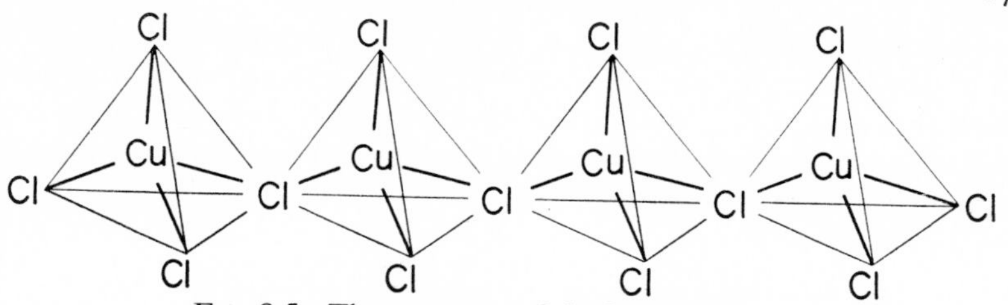

FIG. 8.5 The structure of the $[CuCl_3^{2-}]_n$ ion.

silver and the halogen. The $CuCl_3^{2-}$ ion has an infinite chain structure with $CuCl_4$ tetrahedra joined by bridging chlorines (Fig. 8.5). HgI_2 has an infinite plane structure consisting of HgI_4 tetrahedra sharing corners (Fig. 8.6).

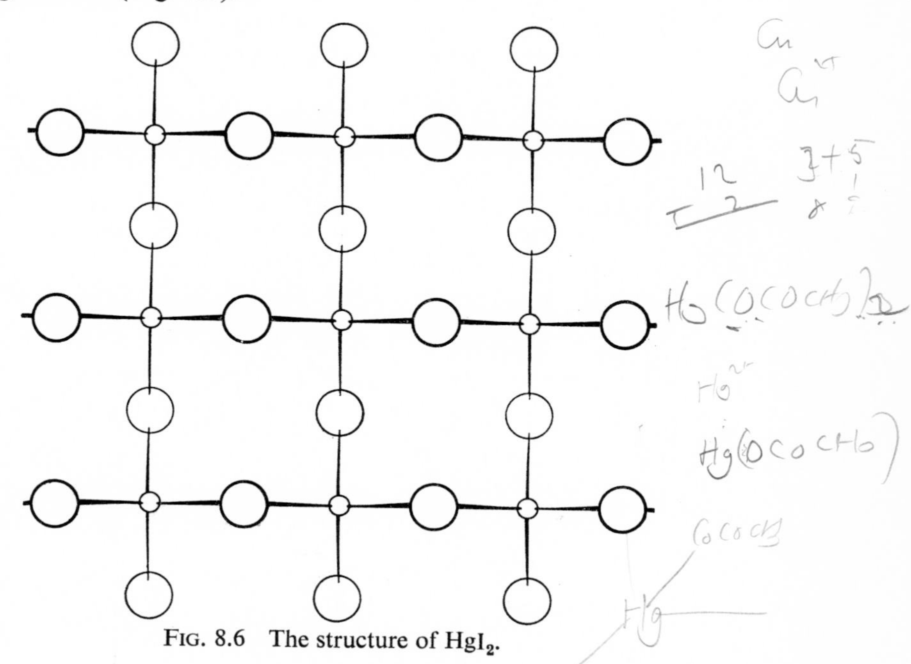

FIG. 8.6 The structure of HgI_2.

8.5 SIX CO-ORDINATION: OCTAHEDRAL AX_6 GEOMETRY

This is the most common co-ordination number and geometry among all the compounds of the transition elements. Because the central core for the vast majority of transition metals carries a charge of at least +3 and because the core is in general large enough to accommodate at least six electron pairs in the surrounding valence shell, these elements can usually hold six electron pairs in their valence shell and thus octahedral six-co-ordination is extremely common. Some examples of simple molecules with this octahedral geometry are given in Table 8.5.

Table 8.5 Octahedral geometry AX_6

d^0	$TiCl_6^{2-}$, TiF_6^{2-}, TaF_6^{2-}, $NbCl_6^-$, $NbCl_5.POCl_3$
	MoF_6, WF_6, WCl_6, $MoO_2Cl_4^{2-}$
d^1	TiF_6^{3-}, VCl_6^{2-}
d^2	VF_6^{3-}, $V(NH_3)_6^{3+}$, CrF_6^{2-}, OsF_6, $ReOCl_3(PEt_2Ph)_2$
d^3	$Cr(NH_3)_6^{3+}$, $Cr(CN)_6^{3-}$, $MnCl_6^{2-}$, $[Re_2OCl_{10}]^{4-}$, IrF_6
d^4	$[Mn(C_2O_4)_3]^{3-}$, PtF_6, $[Ru_2OCl_{10}]^{4-}$
d^5	$Mn(H_2O)_6^{2+}$, $Mn(SCN)_6^{4-}$, $Fe(C_2O_4)_3^{3-}$, $(CoF_6)^{2-}$
d^6	$Fe(H_2O)_6^{2+}$, $Fe(CN)_6^{4-}$, $Cr(CN)_6^{3-}$, NiF_6^{2-}
	$Co(NH_3)_6^{3+}$, $Pt(NH_3)_4Cl_2^{2+}$, $PtCl_6^{2-}$, $PdCl_6^{2-}$
d^7	$Co(NH_3)_6^{2+}$, $Co(NO_2)_6^{4-}$
d^8	$Ni(NH_3)_4(NO_2)_2$, $Nipy_4Cl_2$
d^{10}	$Zn(NH_3)_6^{2+}$

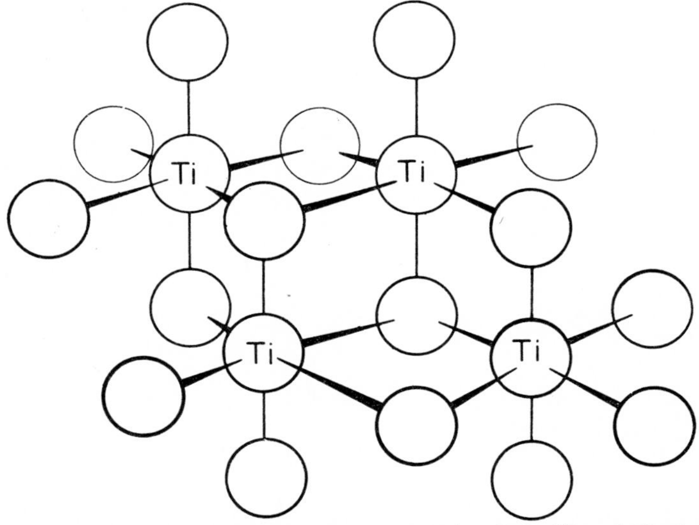

FIG. 8.7 The tetrameric structure of $Ti(OC_2H_5)_4$. Only Ti and O atoms are shown.

$Ti(OC_2H_5)_4$ has an interesting cage structure with octahedral six-co-ordination around each titanium (Fig. 8.7). $TiCl_4.POCl_3$ has the dimeric structure shown in Fig. 8.8 in which titanium has octahedral six-co-ordination. $NbOCl_3$ has a similar structure in which

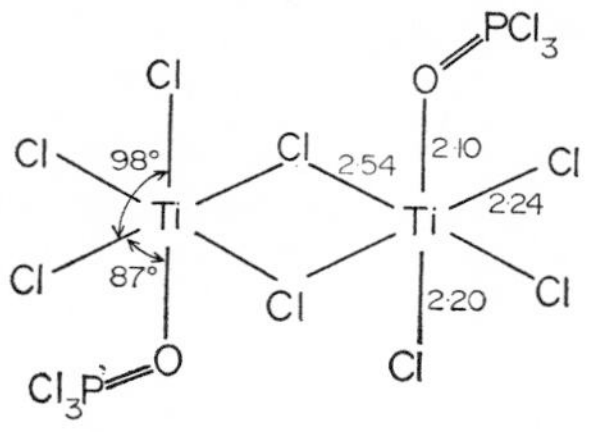

FIG. 8.8 The structure of $(TiCl_4.POCl_3)_2$.

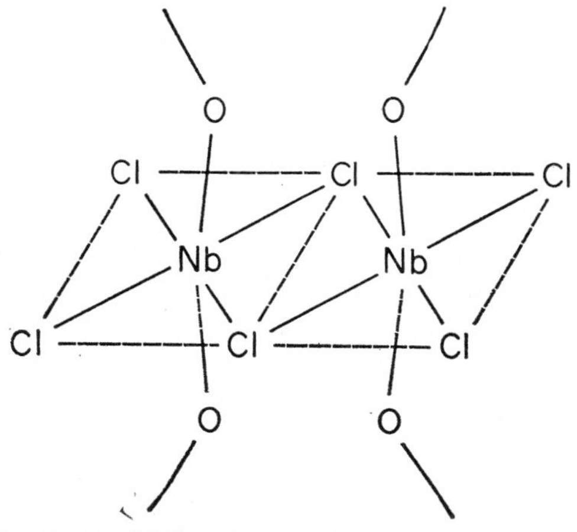

FIG. 8.9 The structure of $NbOCl_3$ in the crystal. The oxygen atoms form bridges between planar Nb_2Cl_6 groups.

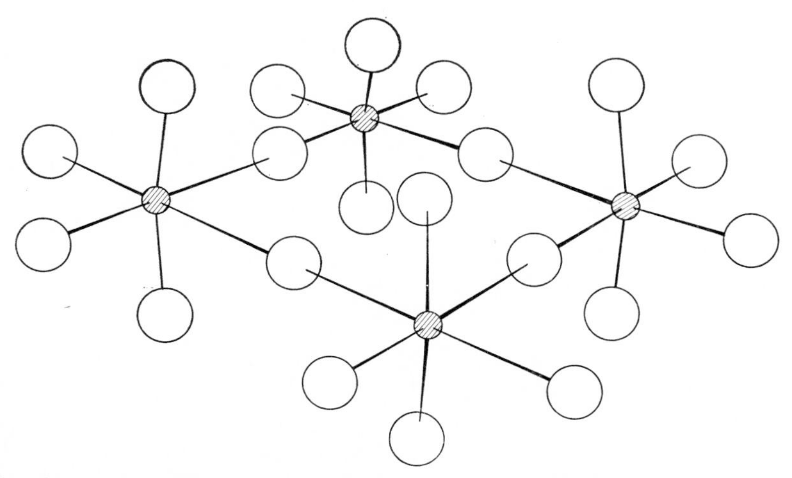

FIG. 8.10 The tetrameric structures of NbF_5, TaF_5, and $MoF_5(d^1)$, and with slight distortion $RuF_5(d^3)$ and $OsF_5(d^3)$.

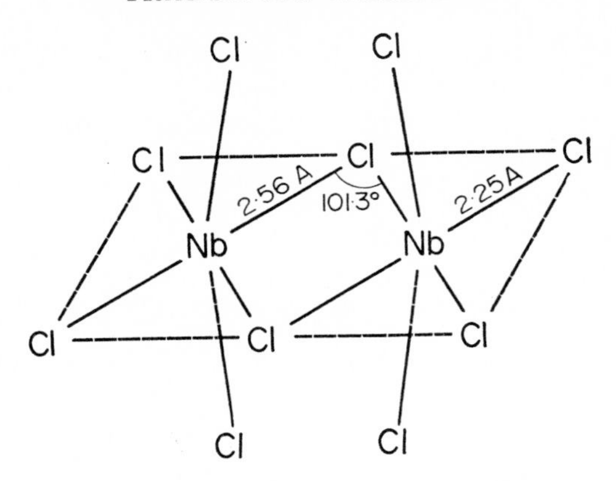

FIG. 8.11 The dimeric structure of $NbCl_5$. The octahedra are distorted as shown.

oxygen atoms bridge between planar chlorine-bridged Nb_2Cl_6 groups (Fig. 8.9). The fluorides NbF_5, TaF_5, $MoF_5(d^1)$, $RuF_5(d^3)$, and $OsF_5(d^3)$ have tetrameric structures in which each metal atom is octahedrally co-ordinated (Fig. 8.10). Niobium pentachloride $NbBr_5$, $TaCl_5$, and $MoCl_5(d^1)$ exist as dimeric molecules with chlorine bridges in which the metal is six-co-ordinated (Fig. 8.11). It is interesting to note that the deviations from the regular octahedral structure around the metal are consistent with the bridge bonds being more polar than the non-bridge bonds and therefore exerting a smaller repulsion. The greater polarity of the bridge bonds results from the fact that the bridging halogen carries a formal positive charge which increases its effective electronegativity. The deviations of the bond angles from 90° are consistent with the

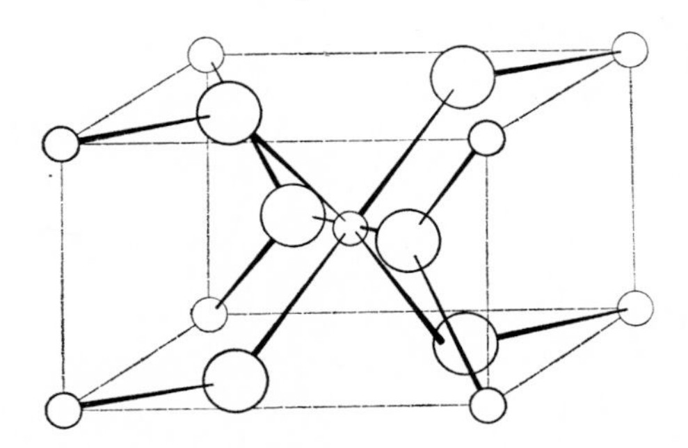

FIG. 8.12 The structure of rutile TiO_2.

oxygen of $POCl_3$ having a considerably higher electronegativity than chlorine. TiO_2 (rutile) has a structure involving octahedral six-co-ordination around titanium and trigonal planar co-ordination around oxygen (Fig. 8.12). Rhenium trioxide (d^1) has an infinite

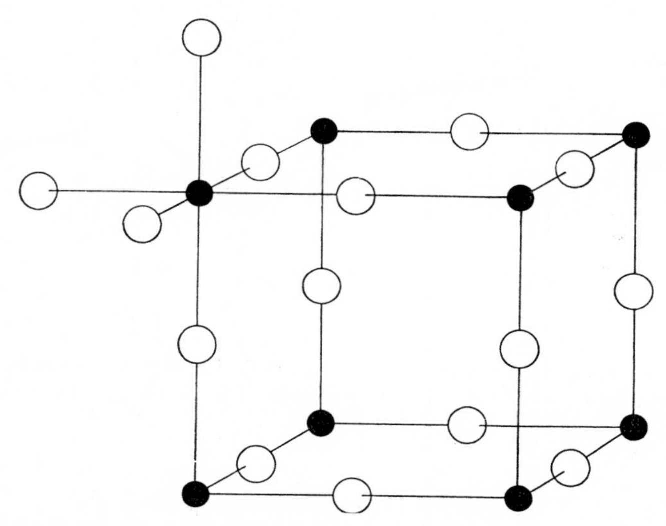

FIG. 8.13 The structure of ReO_3. Each metal atom lies at the centre of an octahedron of oxygen atoms.

lattice structure with octahedral co-ordination around rhenium. $MoF_3(d^3)$, $TaF_3(d^2)$, and $NbF_3(d^2)$ also have this structure (Fig. 8.13).

8.6 SIX-CO-ORDINATION: TETRAGONAL AX_6 GEOMETRY

A prolate ellipsoidal d shell will cause a tetragonal distortion of an octahedral complex such that two axial ligands are found at a greater distance from the central atom than the four equatorial ligands (Fig. 8.14). In the extreme case these two axial ligands may be lost, giving rise to a square planar AX_4 geometry. Tetragonally distorted octahedral geometry has been observed for elements with d^4 and d^9 non-bonding shells (Table 8.6). It may be noted that these are all

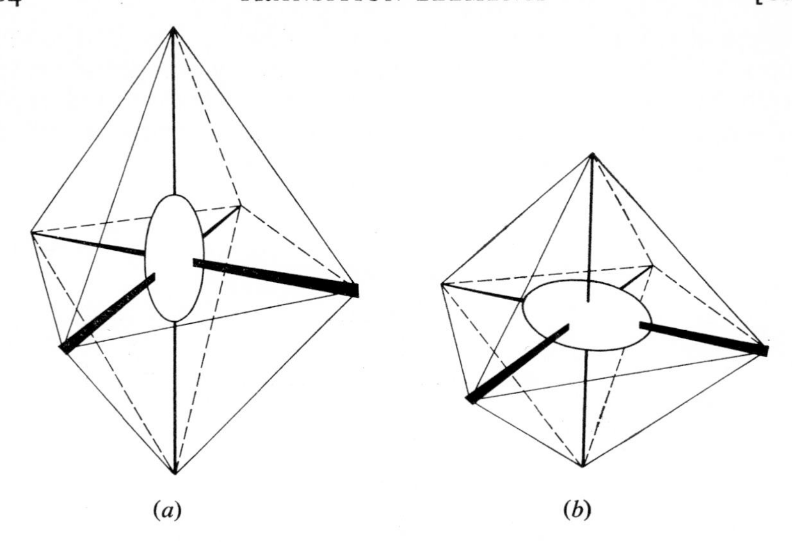

FIG. 8.14 Tetragonal distortion of a octahedral complex caused by a non-spherical (ellipsoidal) d-shell: (a) prolate ellipsoidal d-shell; (b) oblate ellipsoidal d-shell.

Table 8.6 Tetragonal AX_6 geometry

			Bond lengths (Å)	
			Equatorial	Axial
		Elongated Octahedra		
d^4	CrF_2	distorted rutile structure[a]	2·00	2·43
d^4	CrS		2·45	2·88
d^4	$CrCl_2$		2·39	2·90
d^9	CuF_2	distorted rutile structure[a]	1·93	2·27
d^9	$CuCl_2.2H_2O$	Fig. 8.15	2·28	2·95
d^9	$CuF_2.2H_2O$	Fig. 8.15	1·89	2·47
d^9	$Cu(NH_3)_2Br_2$	Fig. 8.16	2·54	3·08
d^9	$CuCl_2$	Fig. 8.16	2·30	2·95
d^9	$CuBr_2$		2·40	3·18
		Flattened Octahedra		
d^4	$KCrF_3$		2·14	2·00
d^9	$KCuF_3$		2·07	1·96
d^9	K_2CuF_4			

[a] For the rutile structure, see Fig. 8.12.

3-dimensional infinite lattice structures. $CuCl_2.2H_2O$ has square planar $CuCl_2.2H_2O$ groups held together by chlorine bridge-bonds giving a distorted octahedral geometry around copper (Fig. 8.15). In $CuCl_2$ for example these are infinite chains formed by square planar $CuCl_4$ groups sharing edges but these chains are so arranged

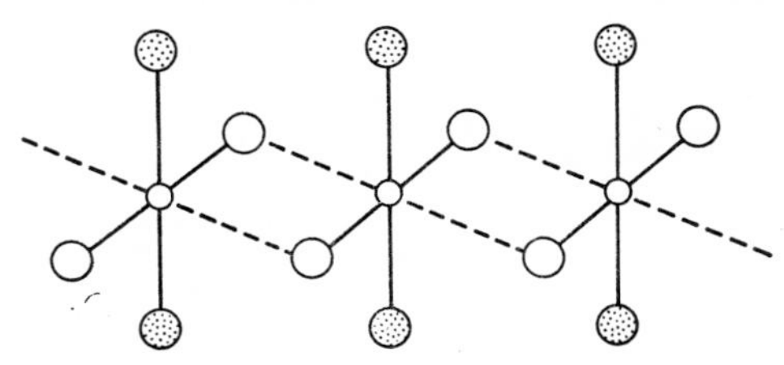

FIG. 8.15 The structure of $CuCl_2.2H_2O$.

that two chlorines from other chains complete a distorted octahedral co-ordination around copper (Fig. 8.16).

If the non-bonding *d* shell has an oblate ellipsoidal shape then the tetragonal distortion of the octahedron will lead to two axial ligands at shorter distances, and four equatorial ligands at longer distances from the central atom (Fig. 8.16). Only a few examples of structures of this type are known, e.g., $KCrF_3$, $KCuF_3$, and K_2CuF_4 (Table 8.6). Each of these structures contain tetragonally distorted octahedral geometry around the metal cation, with four long distances in the equatorial plane and two short distances in the axial direction.

8.7 FOUR CO-ORDINATION: SQUARE PLANAR AX_4 GEOMETRY

Square planar geometry arises from the interaction of a prolate ellipsoidal non-bonding *d* shell with the valence shell and can be regarded as the limiting case of distortion of an octahedral complex where the axial ligands are repelled to a very large distance by the non-bonding *d*-shell or of the distortion of a tetrahedral arrangement of four valence shell electron pairs which are repelled away from the two-fold axis of the tetrahedron and towards a square planar arrangement (Fig. 8.17).

If the distortion is small a rather flattened tetrahedral geometry would be expected. If, on the other hand, the *d* shell has an oblate ellipsoidal shape, then a tetrahedral arrangement will be distorted to an elongated tetrahedron (Fig. 8.17).

Many examples of square planar geometry are found among the

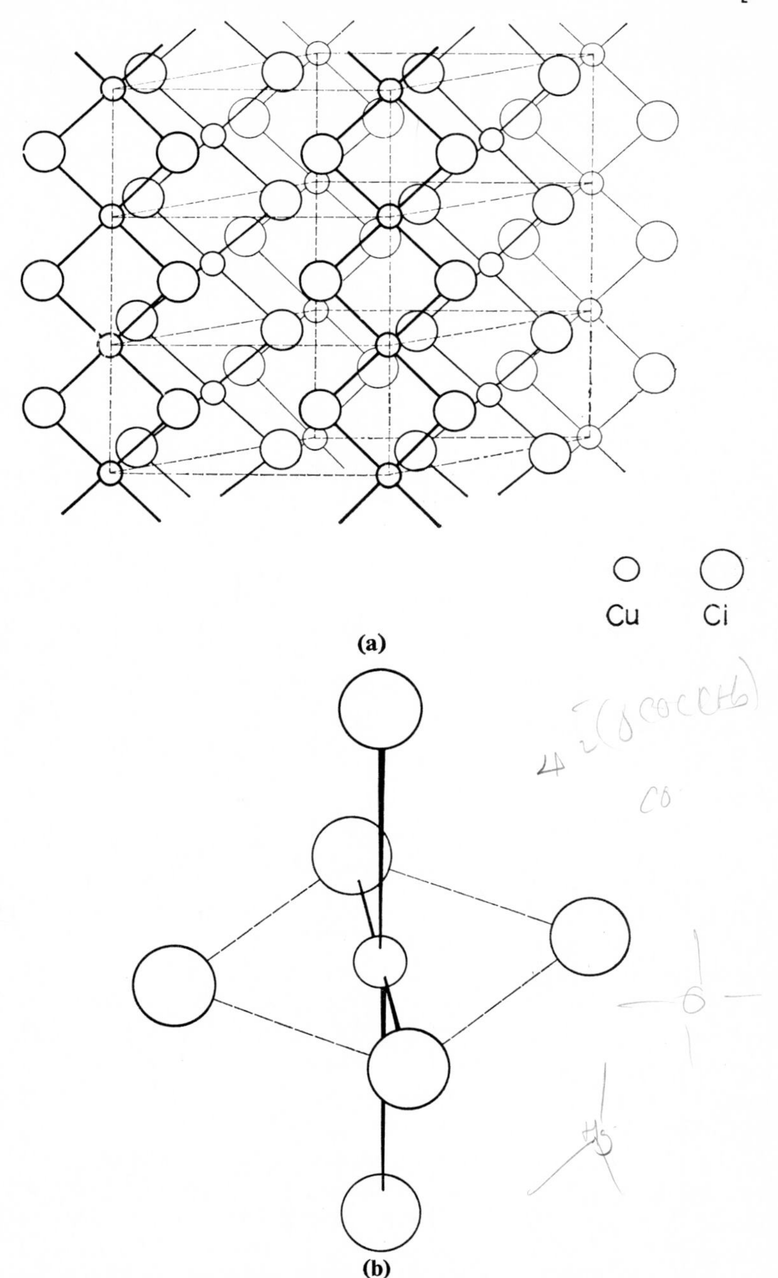

FIG. 8.16 (*a*) The structure of $CuCl_2$; (*b*) Geometry of Cu in $CuCl_2$: ○Cu ◯Cl.

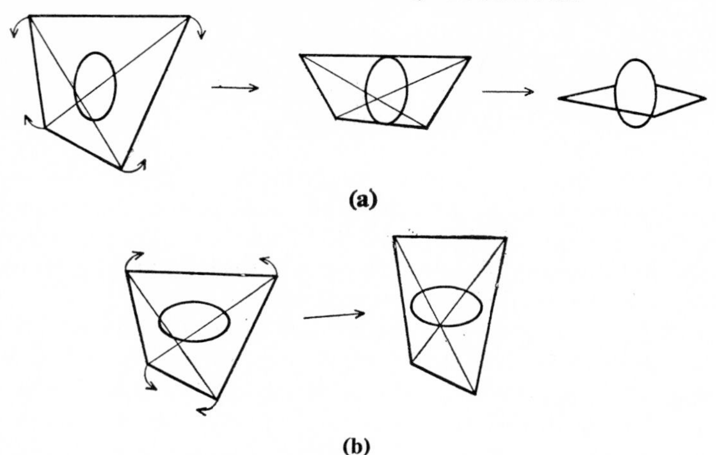

FIG. 8.17 Distortion of a tetrahedral valence shell of four electron pairs by a ellipsoidal *d*-shell: (*a*) prolate ellipsoidal *d*-shell; (*b*) oblate ellipsoidal *d*-shell.

compounds of Ni(II) d^8, Pt (II) d^8, and Pd (II) d^8. Examples are given in Table 8.7.

The only known example of a structure intermediate between the tetrahedral structure determined by interaction between four bonding pairs of electrons and the square planar structure determined by dominant interaction with an ellipsoidal *d*-shell is $CuCl_4^{2-}$ which has the shape of a flattened tetrahedron with bond angles of 120° and 104°.

Table 8.7 Square planar AX_4 geometry

Ni(II)	$Ni(CN)_4^{2-}$
	bis(dimethylglyoximato)nickel(II)
Pd(II)	$PdCl_4^{2-}$ $Pd(NH_3)_4^{+}$ $Pd(NH_3)_2Cl_2$
Pt(II)	$Pt(NH_3)_4^{2+}$ $Pt(NO_2)_4^{2-}$ $Pt(NH_3)_2Cl_2$ $PtCl_4^{2-}$ $Pt(CN)_4^{2-}$ $[P(C_2H_5)_3]_2PtCl_2$ $[P(C_2H_5)_3]_2PtHBr$

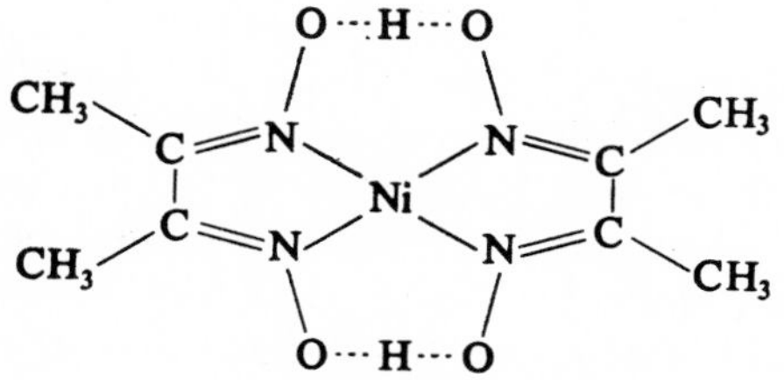

8.8 FIVE-CO-ORDINATION: AX_5 TRIGONAL BIPYRAMID AND TETRAGONAL PYRAMID GEOMETRY

For d^0, d^5, and d^{10} non-bonding shells we expect the regular trigonal bipyramid geometry with axial bonds longer than equatorial bonds. Some examples are given in Table 8.8. The data for the niobium and tantalum halides are not accurate enough to show any differences in the axial and equatorial bond lengths.

In $CdCl_5^{3-}$ the axial and equatorial bond lengths which have been reported to be 2·53 and 2·56 Å respectively do not show the expected difference; indeed if the apparent difference is significant it is in the wrong direction. In the other molecules the ligands are either chelating or bridging and a variety of factors may therefore be influencing the bond lengths.

In crystalline $KVO_3.H_2O$ the vanadium is surrounded by five oxygen atoms with a distorted trigonal bipyramid arrangement. Each

Table 8.8 AX_5 trigonal bipyramid molecules of transition elements with d^0, d^5, and d^{10} configurations

d^0	$NbCl_5$, $NbBr_5$, $TaCl_5$, $TaBr_5$, $KVO_3.H_2O$, V_2O_5
d^5	$[Mn(Sal—Me)_2]_2$
d^{10}	$[CdCl_5^{3-}]_2Zn(acac)_2H_2O$ $[Zn(Sal—Me)_2]_2$

(Sal—Me) = N-methylsalicylaldiminate (see Table 8.2)
acac = acetylacetonate

trigonal bipyramid shares two edges with neighbouring trigonal bipyramids to form an infinite chain as shown in Fig. 8.18. If we assume that all the bonds are covalent bonds, then the atoms carry the formal charges shown. The V—O bonds have lengths of 1·63 Å while the remaining three have lengths in the range 1·93–1·99 Å. These three long bonds are to oxygen atoms that have formal positive charges and are forming a total of three bonds to vanadium atoms, and they must therefore be essentially single bonds. The two short bonds are to oxygen atoms that are not bonded to other vanadium atoms, and it is reasonable therefore to suppose that they have some double-bond character, as indeed their length suggests. These two multiple bonds cause the observed distortions of the trigonal bipyramid, decreasing the angle between the three single bonds from the ideal value of 90° to 74°. This effect is accentu-

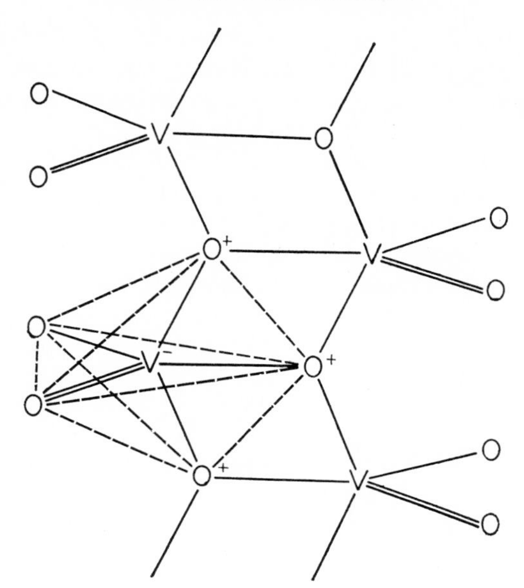

FIG. 8.18 The structure of $KVO_3.H_2O$.

ated by the fact that these single bonds are to highly electronegative O^+ atoms which cause the bonding pairs to be pulled away from the vanadium, further reducing their mutual repulsions.

The structure of V_2O_5 (Fig. 8.19) is clearly similar to that of the

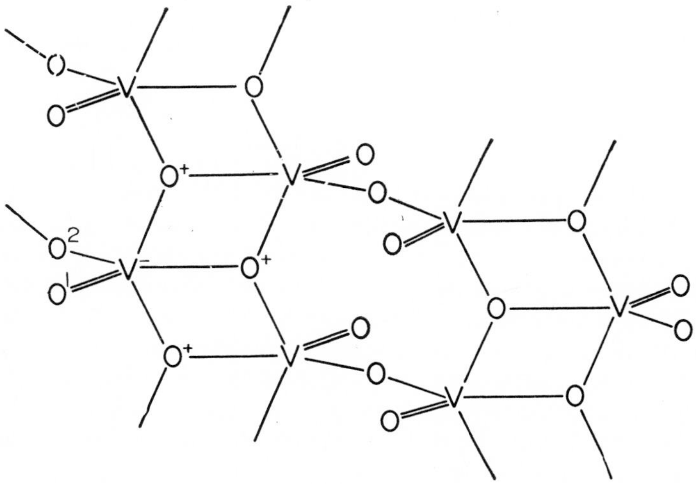

FIG. 8.19 The structure of V_2O_5.

$(VO_3^-)_n$ ion. It contains the same infinite chain of trigonal bipyramids sharing edges, but these chains are joined together by sharing another oxygen atom at a corner to give a three-dimensional complex. Thus there are three different types of oxygen atom surrounding the vanadium. O^1 is bonded to only one vanadium, and the bond VO^1 which has a length of 1·54 Å is presumably a double bond. This length is close to the VO distance in $VOCl_3$. The oxygen atom O^2 is linked to two vanadium atoms by bonds which it is reasonable to assume have some double-bond character. They have a length of

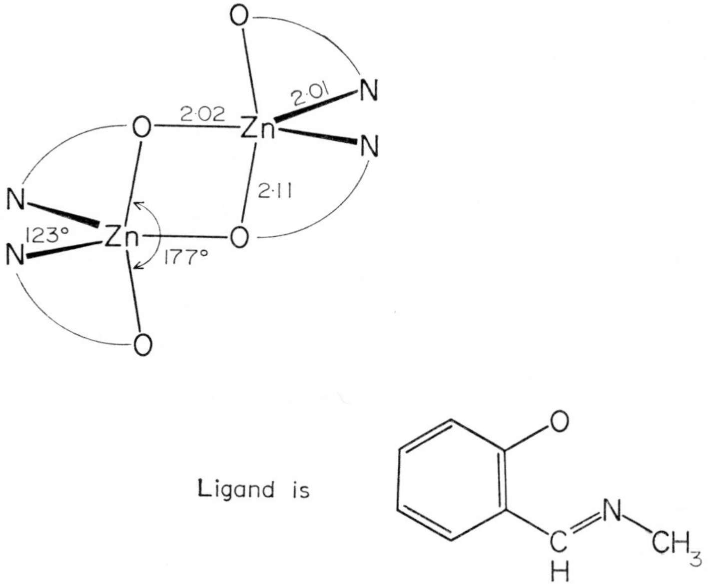

FIG. 8.20 The structure of the N-methylsalicaldimine complex of zinc (II).

1·77 Å and the angle at the oxygen atom is 125°, which also indicates some double-bond character. The other three bonds may be regarded as single bonds to the very electronegative O^+ atoms and they have lengths of 1·85–2·02 Å. The dimeric *N*-methylsalicaldimine complexes of manganese (II) and zinc (II) and also Co (II) are isomorphous and isostructural. The zinc complex has been shown to occur as dimers (Fig. 8.20) with each zinc atom in a somewhat distorted trigonal pyramid arrangement. As expected, the axial metal–oxygen bond length is greater than the equatorial bond lengths and the equatorial bond angle is approximately 120°.

There appears to be one known exception to the prediction that five-co-ordinated molecules of the transition elements with spherical *d* shells will have trigonal bipyramid structures and that is *NN″*-disalicylidene ethylenediaminezinc monohydrate which has a tetragonal pyramid structure with the zinc atom lying 0·34 Å above the base of the square pyramid and with the water molecule in the axial position at 2·13 Å from the zinc compared with 1·94 Å for the other metal–oxygen distances (Fig. 8.21). This stereochemistry is evidently forced by the ligand as a model shows that it is not possible to have all four of the co-ordinating groups of the tetradentate ligand at four of the vertices of a trigonal bipyramid. As in all tetragonal pyramid molecules in which there is not a lone-pair of electrons occupying the sixth octahedral position, the metal is above the plane of the base, the

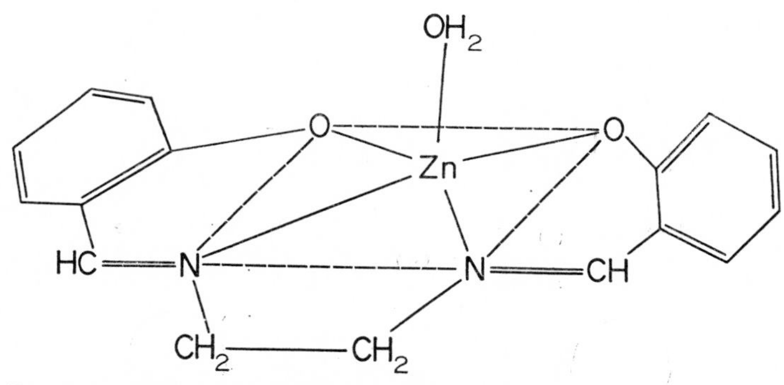

FIG. 8.21 NN-disalicylindene ethylenediamine zinc monohydrate.

ideal bond angle between the axial ligand and the ligands in the base being about 100°. Figure 8.22 shows that the interaction between a prolate ellipsoidal *d* shell and the valence shell will cause the valence shell electron pairs to avoid the ends of the ellipsoid and thus the trigonal bipyramid will be destabilized with respect to the square pyramid which has only one electron pair in such a position. Thus when there is a strong interaction between the valence shell and the *d* shell, the square pyramidal configuration for five electron pairs will be preferred to the trigonal bipyramid arrangement. In the square pyramidal configuration the additional repulsion exerted on the axial electron pair by the *d* shell will lead to the axial bond being longer than the equatorial bonds. The interaction of an oblate ellipsoidal *d* shell would however further stabilize the trigonal bipyramid with respect to the square pyramid, and would be expected to reduce the difference in length between the axial and equatorial bonds so that this might become small or even in the opposite direction from that observed for the main group elements. Unfortunately, the factors that determine whether or not the *d* shell should be regarded as having a prolate or an oblate shape are not clear. However, one can predict that square pyramid structures with longer axial than equatorial bonds are likely to be observed and that if the trigonal bipyramid structure is found there may be only a very small difference between the lengths of the axial and equatorial

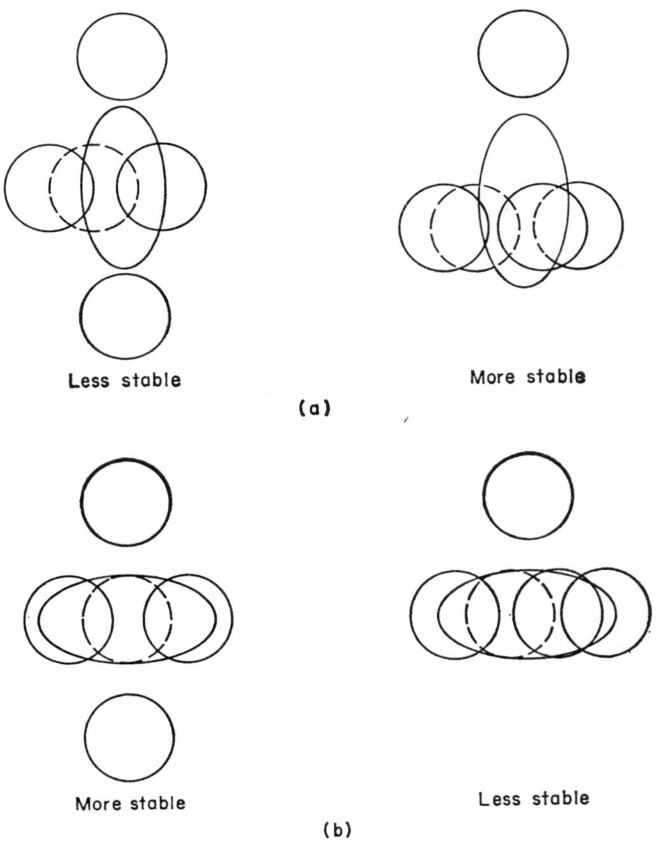

FIG. 8.22 Effect of a non-spherical (ellipsoidal) *d*-shell on the arrangement of five electron pairs in a valence shell: (*a*) prolate ellipsoidal *d*-shell; (*b*) oblate ellipsoidal *d*-shell.

bonds: and the axial bonds may even be shorter than the equatorial bond.

Table 8.9 gives examples of five-co-ordinated molecules with non-spherical *d* shells which have been found to have trigonal bipyramid

Table 8.9 Trigonal bipyramidal molecules of the transition element with incomplete (non-spherical) inner *d* shells

			Bond lengths	
			Axial	Equatorial
$[Co(Sal—Me)_2]_2$	d^7	Fig. 8.19		
$[Pt(QAs)I]^+$	d^8	Fig. 8.23		
$Pt(SnCl_3)_5^{3-}$	d^8		2·54	2·54
$Co(CNCH_3)_5^+$	d^8		1·87	1·87
$RhH(CO)[P(C_6H_5)_3]_3$	d^8	Fig. 8.24		
$CuCl_5^{3-}$	d^9		2·30	2·39
$[Cu(bipy)_2I]^+$	d^9		2·02	2·02

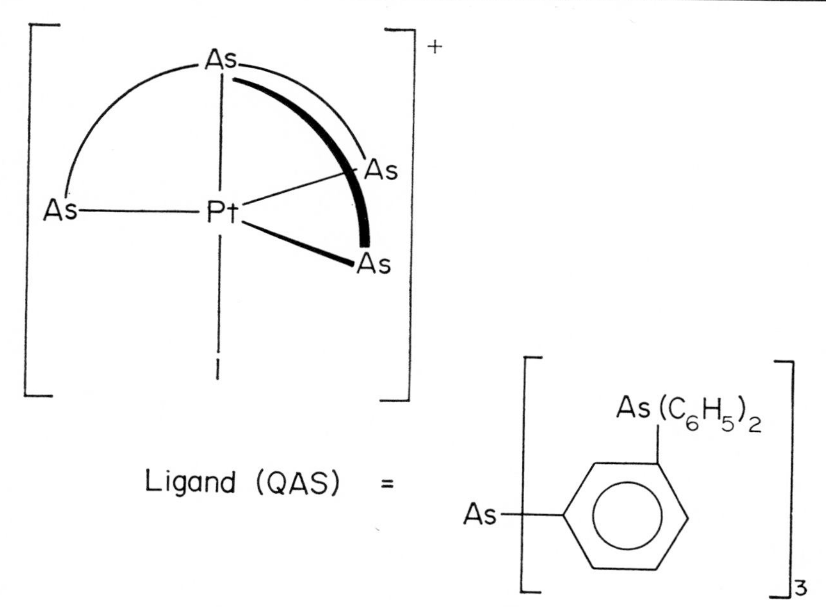

FIG. 8.23 The structure of $Pt(QAS)I^+$.

structures. In $Pt(SnCl_3)_5^{3-}$ and $Co(CNCH_3)_5^+$, there appears to be a negligible difference in the axial and equatorial bond lengths, and in the case of $CuCl_5^{3-}$ the axial bonds appear to be slightly shorter than the equatorial bonds. In the triphosphine complex of rhodium carbonyl hydride (Fig. 8.24) the phosphine ligands are in the equatorial positions, but are bent somewhat towards the hydrogen which is consistent with our earlier postulate that the size of an X—H bond-pair is smaller than that of other bond-pairs on the same central

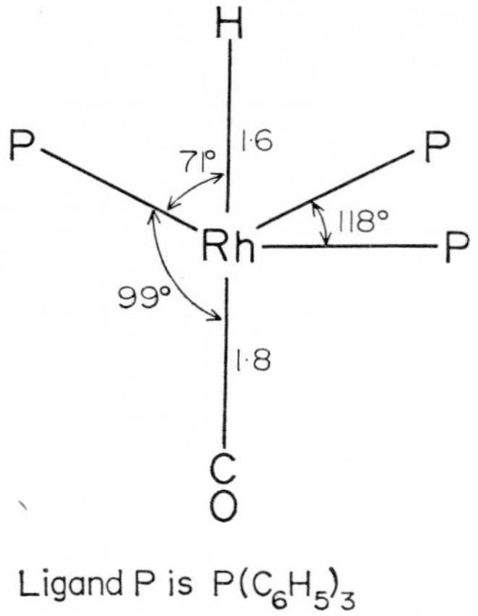

Ligand P is $P(C_6H_5)_3$

FIG. 8.24 Tris(triphenylphosphine)rhodium carbonyl hydride.

atom. The *N*-methylsalicylaldimine complex of cobalt (II) has the same structure as the corresponding zinc complex (Fig. 8.20).

Table 8.10 gives examples of five-co-ordinated molecules with non-spherical *d*-shells which have been found to have tetragonal pyramid

Table 8.10 Square pyramidal molecules of the transition elements with incomplete (non-spherical) *d* shells

		Bond lengths		Displacement of metal atom above base of	
		Axial	Equatorial	square pyramid	Figure
$VO(acac)_2$	d^1	1·56	1·97		8·25
$RuCl_2(PPh_3)_3$	d^6	2·39	2·23		8·26
$PdBr_2(PPh_3)_3$	d^8	2·93	2·52		8·27
$TriarsNiBr_2$	d^8	2·69	2·37		8·28
$Ni(DEAS)_2$	d^8	1·93	2·04	0·36	8·29
$Cu(DMG)_2$	d^9	2·43	1·94		8·30
$Cu[S_2CN(CH_3)_3]_2$	d^9	2·71	2·32	0·4	8·31
$Cu(salicyl\text{-}en)_2$	d^9	2·41	2·01		8·32
$IrCl(CO)(SO_2)[P(C_6H_5)_3]_2$	d^8			0·21	8·33

structures. In almost every case the metal atom lies a significant distance above the base of the pyramid and the axial bond is longer than the bonds in the base of the pyramid. Exceptions are the $Ni(DEAS)_2$ complex which has a slightly shorter axial than base-bonds and for which there is no obvious explanation and bis(acetyl-

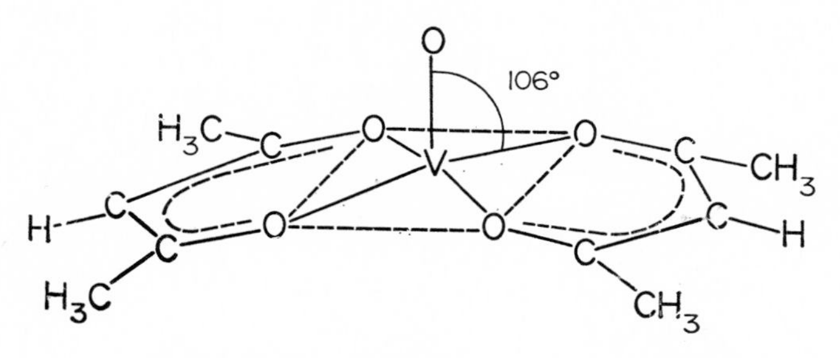

FIG. 8.25 Bis(acetylacetonato)oxavanadium(IV).

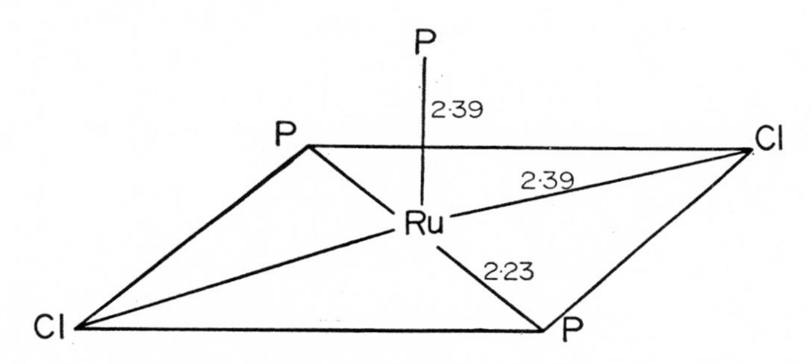

FIG. 8.26 Dichlorotris(triphenylphosphine)ruthenium(II).

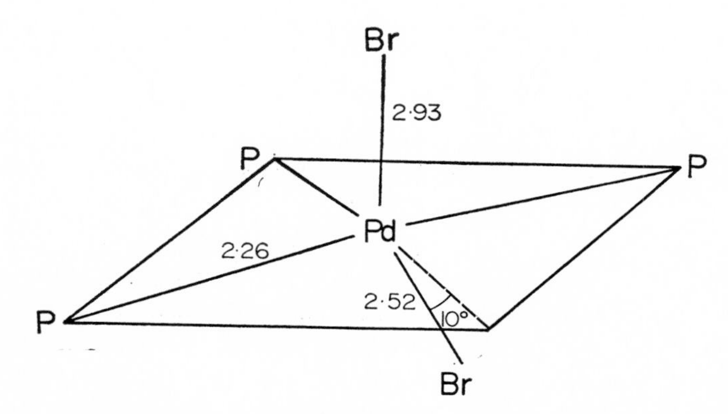

FIG. 8.27 Dibromotris (triphenylphosphine) palladium(II).

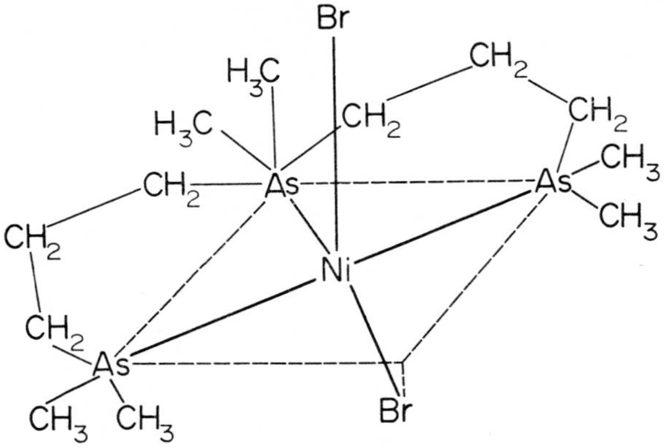

FIG. 8.28 Triarsine complex of nickel dibromide.

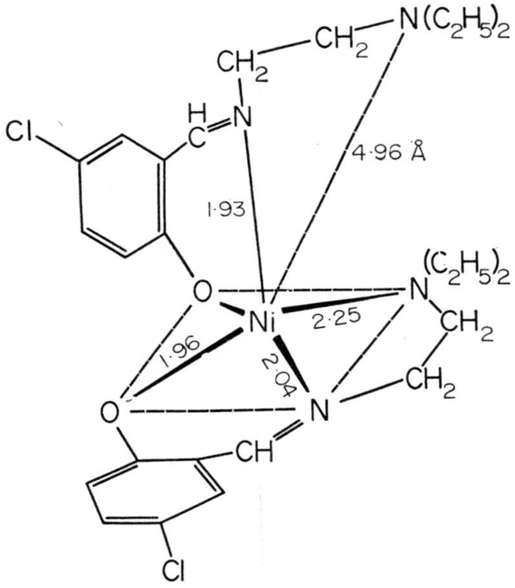

FIG. 8.29 The 2 : 1 N-β-diethylamine-ethyl-5-chlorosalicyaladime complex of Ni(II).

acetonato)oxovanadium(IV) which has a very short axial bond. Although this latter complex has one electron in the *d* shell this would not be expected to cause any appreciable distortion of a regular trigonal bipyramid arrangement of the bonding electron pairs. The short VO bond is, however, presumably a multiple bond with a bond order of at least two. Such a multiple bond would always occupy an

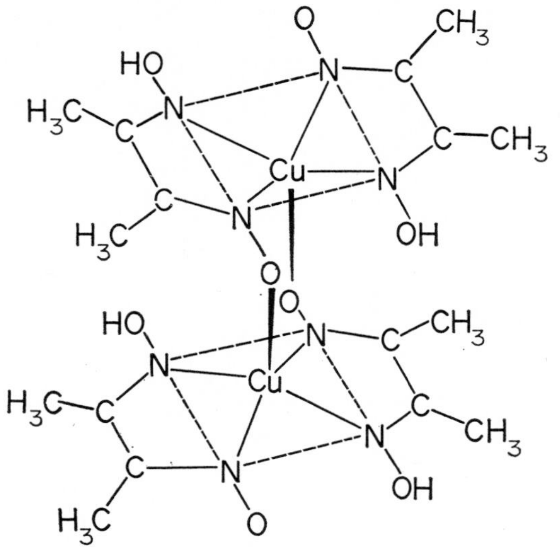

FIG. 8.30 Bis(dimethylgyloximato)copper(II).

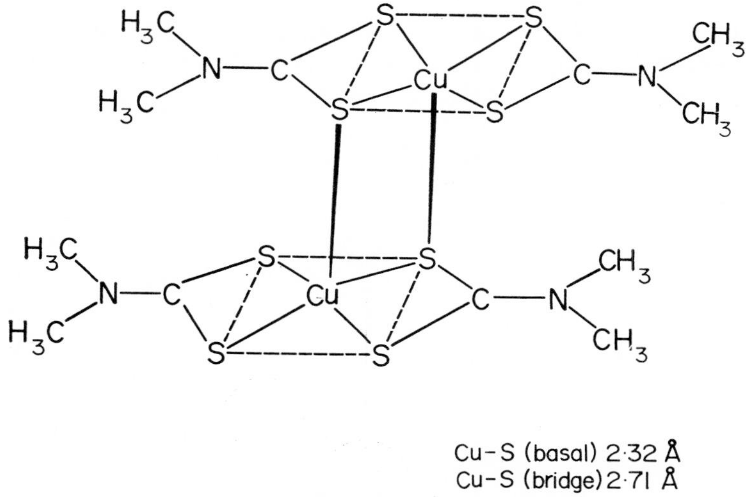

Cu–S (basal) 2·32 Å
Cu–S (bridge) 2·71 Å

FIG. 8.31 Bis(dithiocarbamato)copper(II).

equatorial position of a trigonal bipyramid and would distort the trigonal bipyramid towards the tetragonal pyramid structure (cf. SOF_4, p. 157, and it seems likely that a strong distortion of this kind occurs in this case. Moreover the single *d* electron may at least partly occupy the vacant sixth octahedral position. The geometry of

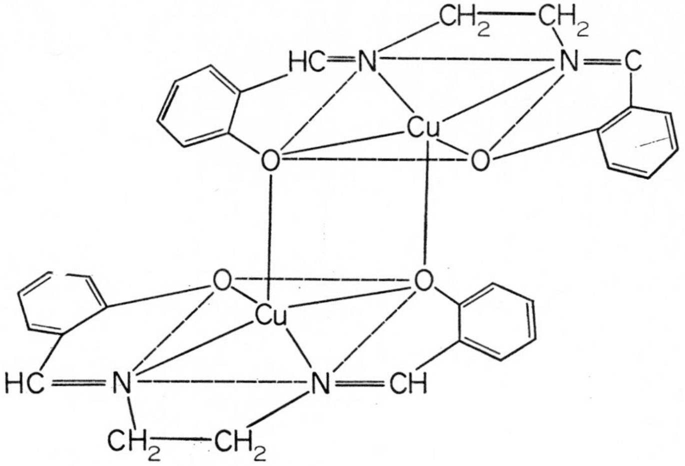

FIG. 8.32 The dimeric molecule of NN-disalicylidene-ethylenediamine copper.

the SO_2 group in the complex $Cl(CO)(SO_2)Ir]P(C_6H_5)_3]_2$ is interesting as it clearly shows the stereochemical effect of the lone-pair on sulphur completing an approximately tetrahedral arrangement of two double bonds, one single bond and a lone-pair around sulphur (Fig. 8.33).

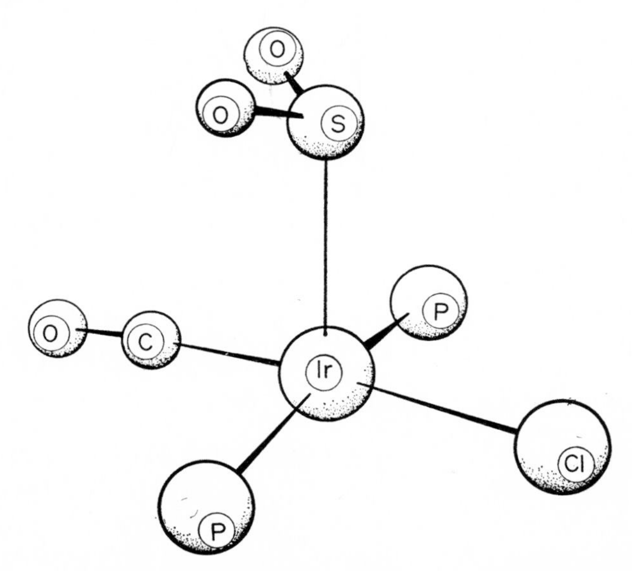

FIG. 8.33 The co-ordination around iridium in $IrCl(CO)(SO_2)[P(C_6H_5)_3]_2$.

8.9 METAL CARBONYLS

Many of the transition metals form metal carbonyls including simple carbonyls such as $Ni(CO)_4$ and $Fe(CO)_5$ as well as more complex polynuclear compounds containing several metal atoms and compounds containing other ligands in addition to the carbonyl group. The simplest formulation of these compounds is based on the metal in a zero oxidation state, each carbonyl group forming one bond to the metal by using the lone-pair of electrons on carbon. However, the metal–carbon distances are always quite short, and it

$$\begin{array}{c} O^{+} \\ \| \! | \\ C \\ | \\ {}^{+}O{\equiv}C{-}Ni^{4-}{-}C{\equiv}O_{+} \\ | \\ C \\ \| \! | \\ O_{+} \end{array} \qquad\qquad \begin{array}{c} O \\ \| \\ C \\ \| \\ O{=}C{=}Ni{=}C{=}O \\ \| \\ C \\ \| \\ O \end{array}$$

(1) **(2)**

is generally accepted that the formally non-bonding *d* electrons on the metal are involved in multiple bond formation with the carbonyl group, leaving an effectively empty (d^0) non-bonding *d*-shell, i.e. $Ni(CO_4)$ is better represented by structure (2) than by (1). Thus the geometry of these molecules may be predicted using the simple rules for d^0 compounds, i.e., we expect to find tetrahedral $M(CO)_4$ molecules, trigonal bipyramidal $M(CO)_5$ molecules, and octahedral $M(CO)_6$ molecules, and this is indeed the case as shown in Table 8.11.

The only unexpected feature of the data in this table is that the equatorial bond length in $Fe(CO)_5$ (3) is slightly longer than the axial

Table 8.11 Structures of some mononuclear carbonyls

Molecule	Shape	M—C Bond length (Å)	
$Ni(CO)_4$	Tetrahedron	1·84	
$Cr(CO)_6$	Octahedron	1·92	
$Mo(CO_6)$	Octahedron	2·06	
$W(CO)_6$	Octahedron	2·06	
		Equatorial	Axial
$Fe(CO)_5$	Trigonal bipyramid	1·84	1·80
$Co(SiCl_3)(CO)_4$	Trigonal bipyramid	1·76	1·80

bond length, whereas for a d^0 system the reverse situation is predicted. It should be noted however that iron has only eight d electrons and therefore does not have sufficient electrons to form a full double bond to each carbonyl group as is indicated by the single valence bond. As the axial and equatorial positions of a trigonal bipyramid

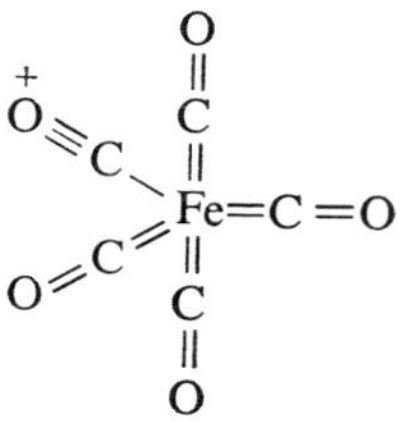

(3)

are not equivalent there is no reason why the extent of double bonding should be the same in both the equatorial and axial directions, and it would appear that it is somewhat greater in the axial than in the equatorial directions. The structure of $Co(SiCl_3)(CO)_4$ is, however, in accordance with expectation (Fig. 8.34). The electro-

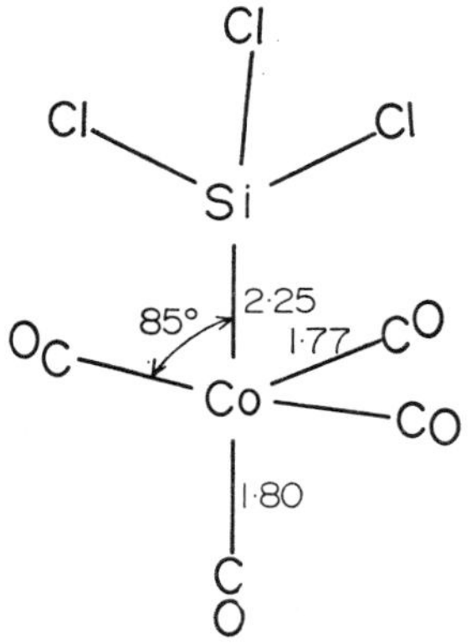

FIG. 8.34 The structure of $Co(SiCl_3)(CO)_4$.

negative $SiCl_3$ group is in an axial position and the axial CO bond length is greater than the equatorial bond length. Moreover the Co atom is displaced below the plane of the equatorial CO groups by 0·15 Å because of the greater repulsion exerted by the Co—C axial bond than by the Co—Si single bond.

The molecule $Zn[Co(CO)_4]_2$ has an interesting structure with a linear Co—Zn—Co arrangement of the metal atoms and a trigonal

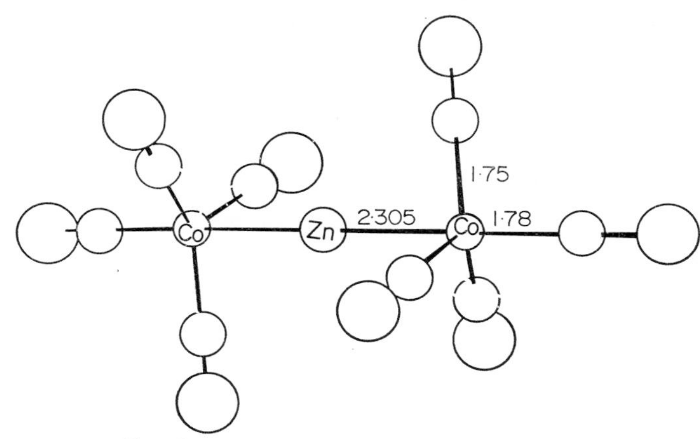

FIG. 8.35 The structure of $Zn[Co(CO)_4]_2$.

bipyramid geometry around each cobalt (Fig. 8.35). Again the axial Co—C bond length of 1·78 Å is slightly greater than the equatorial bond length of 1·75 Å.

The structures of several binuclear carbonyls are shown in Fig. 8.36. In each case there is a metal–metal bond and there may also

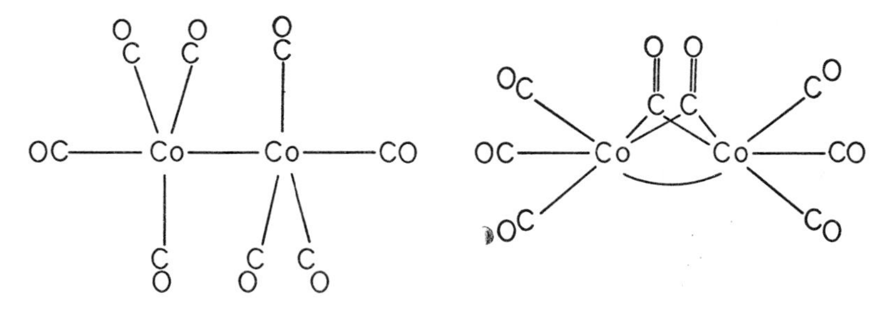

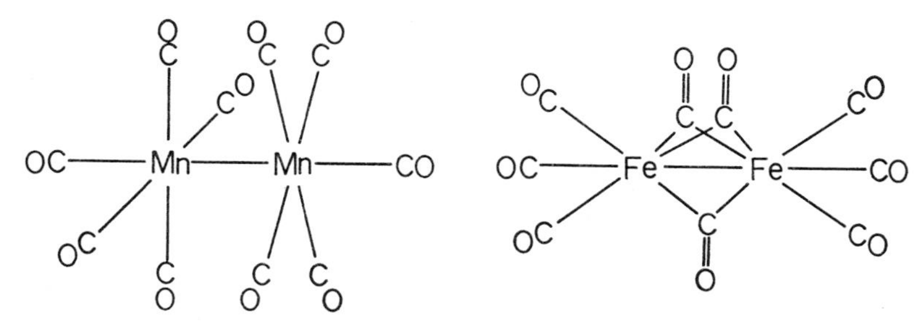

FIG. 8.36 Structures of some binuclear metal carbonyls.

be bridging carbonyl groups which form only single bonds to each metal. Any remaining normally non-bonding *d*-electrons on the metal are assigned to double bonds with the terminal carbonyl groups so that each metal atom effectively has a d^0 configuration. The binuclear cobalt carbonyl is found in two forms, the non-bridged form in the crystalline state and the carbonyl bridged form in solution. In the former there is trigonal bipyramid geometry around each cobalt, while in the latter each cobalt has octahedral geometry counting the bent metal–metal bond. In $Mn_2(CO)_{10}$ each manganese has an octahedral arrangement of bonds, while in the binuclear iron carbonyl there is a seven-co-ordinate 1 : 3 : 3 arrangement around each iron, counting the iron–iron bond—this is based on an octahedral arrangement of the carbonyl ligands which is somewhat deformed by the presence of the metal–metal bond.

In the trinuclear and polynuclear carbonyls co-ordination numbers of greater than six around the metal atom are often encountered and, moreover, the bond angles may be considerably deformed by the formation of polynuclear clusters of metal-atoms held together by metal–metal bonds, and therefore their structures cannot be discussed in a simple manner. Some examples of molecules containing metal–metal bonds are discussed in Section 8.11.

8.10 PEROXY COMPOUNDS

Several metals form a number of interesting peroxy compounds in which a peroxy group functions as a bidentate ligand. However,

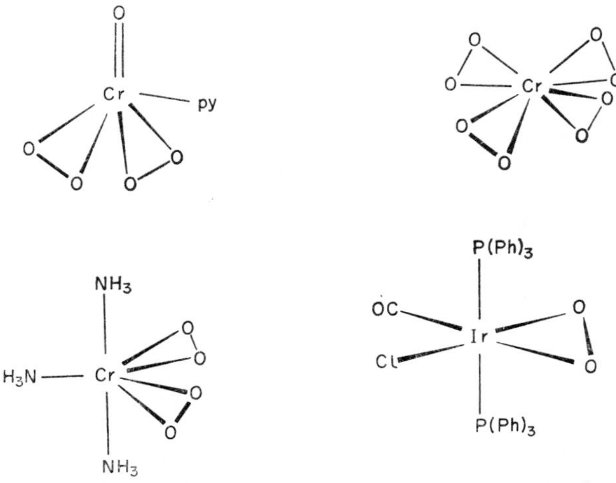

FIG. 8.37 Structures of some metal peroxy compounds.

because of the short O—O distance of approximately 1·4 Å, the two electron pairs of the bonds to the metal are held quite close together and function rather like a double-bond pair. If this is assumed to be the case then the structures of these peroxy compounds follow the usual rules (Fig. 8.37). Thus $CrO(O_2)_2py$ can be regarded as having tetrahedral geometry. CrO_8^{3-}, which in fact has bisdisphenoid structure, can also be regarded as a flattened tetrahedron, the repulsion between the double bonded O_2 groups increasing the angle between them from the tetrahedral angle to the observed angle of 140°. In $Cr(O_2)_2(NH_3)_3$ there is a trigonal bipyramid arrangement of the ligands around the central chromium in which the 'doubly bonded' O_2 groups occupy equatorial positions as expected. A similar trigonal bipyramid geometry has been found for the complexes $Ir(O_2)Cl(CO)(PPh_3)_2$ and $IrO_2I(CO)(PPh_3)_2$.

8.11 METAL–METAL BONDS AND CLUSTER COMPOUNDS

There is, in general, nothing unusual about the nature of the bonds between metal atoms in molecules, although these bonds were originally thought to be rare and rather unusual. Such bonds may consist of one, two, or three shared electron pairs as for non-metallic elements. One unusual feature of metal–metal bonds is that even four shared electron pairs is apparently possible as there appears to be a quadruple bond in the $Re_2Cl_8^{2-}$ ion which has the structure shown in Fig. 8.38. Each rhenium has seven electrons in the neutral atom and may be regarded as acquiring one additional electron on the formation of the anion. Of these eight electrons, four are used in the formation of bonds to chlorine, leaving four unused on each rhenium. Since the compound is diamagnetic and has a very short rhenium–rhenium bond of 2·24 Å it has been proposed by Cotton that there is a quadruple bond between the two metal atoms. This would then consist of four pairs of electrons arranged in the form of a square normal to the Re–Re direction as shown in Fig. 8.38. This also accounts for the observed stereochemistry of the molecule, which is at first sight unusual, as the chlorines at the two ends of the molecule adopt an eclipsed arrangement. However, the four electron pairs of the quadruple bond complete an approximately square antiprism arrangement of electron pairs around each rhenium atom, and as the four pairs of the quadruple bond are therefore necessarily staggered with respect to both sets of four pairs of chlorine bonding electrons it follows that these must be eclipsed. The trimeric

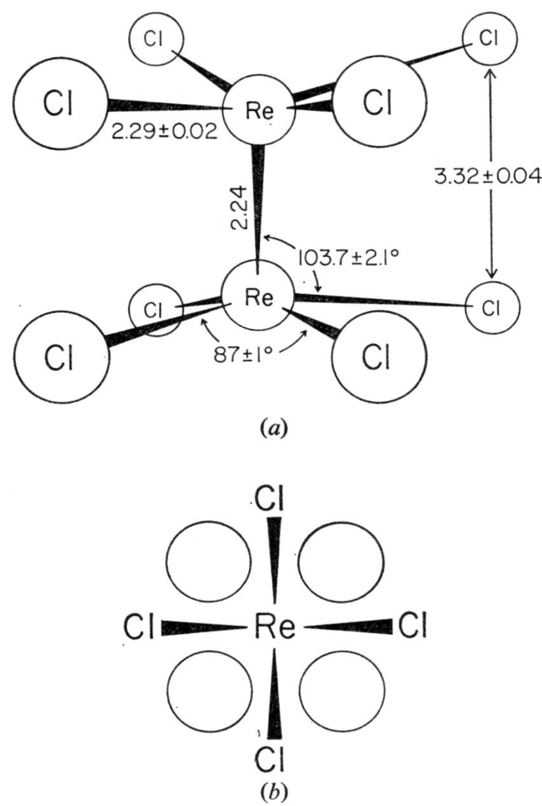

FIG. 8.38 (*a*) The structure of the $Re_2Cl_8^{2-}$ anion; (*b*) view down the Re-Re axis of the $Re_2Cl_8^{2-}$ anion showing the staggered arrangement of the four Re-Cl bonds and the four electron pairs of the quadruple bond.

$Re_3Cl_{12}^{3-}$ anion is also known and has a triangular group of rhenium atoms as shown in Fig. 8.39. The bonding in this compound can be described in a similar manner, but in this case each rhenium atom forms two double bonds to its neighbouring atoms and five bonds to chlorines, of which two are bridging, resulting in the tricapped trigonal prism arrangement of nine electron pairs around each rhenium. This same molecular geometry is also found in $ReCl_3$, in which trimeric Re_3Cl_9 groups are linked by chlorine bridges, and in the ions $Re_3Cl_{11}^{2-}$ and $Re_3Cl_{10}^{-}$, which are formed from $Re_3Cl_{12}^{2-}$ by loss of one or two of the terminal chlorine atoms respectively as chloride ions, leaving a square antiprism arrangement of the remaining eight bonds around each rhenium.

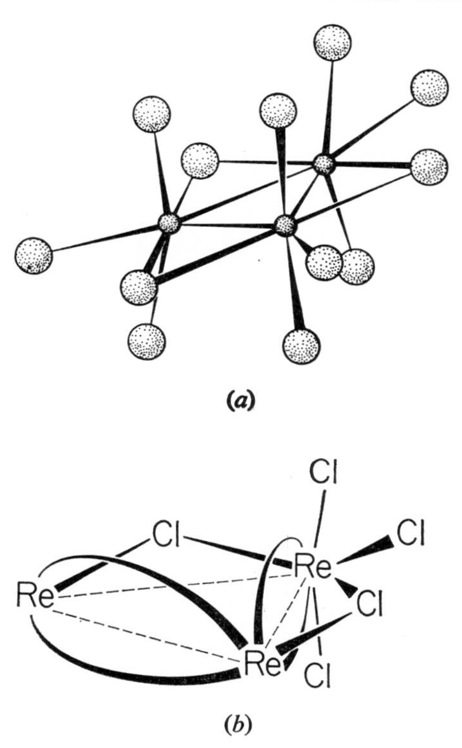

FIG. 8.39 (*a*) The structure of the $Re_3Cl_{12}^{3-}$ anion; (*b*) bonding in $Re_3C_{12}^{3-}$. For clarity all the Re-Cl bonds are shown only for one Re atom and one Re-Re double bond is omitted.

Octahedral clusters of metal atoms are found in the ions $Mo_6Cl_8^{4+}$ and $Ta_6Cl_{12}^{2+}$ (Fig. 8.40). On counting the electrons in $Mo_6Cl_8^{4+}$ one finds that of the thirty-six electrons of the four molybdenum atoms eight must be used to bond the chlorines and four are lost to give the positive charge, leaving a total of twenty-four electrons, which is just the required number to form twelve bonds along the edges of the octahedron. Thus each molybdenum appears to form four metal–metal bonds along the edges of the octahedron and an approximately square antiprism arrangement of eight electron pairs around each molybdenum is completed by the bonds to four chlorines which lie above the faces of the octahedron and bridge to other chlorines (Fig. 8.41). A similar electron count for the $Ta_6Cl_{12}^{2+}$ cation shows that there are sixteen electrons available for metal-bonding in the Ta_6 octahedron, and these can be regarded as forming

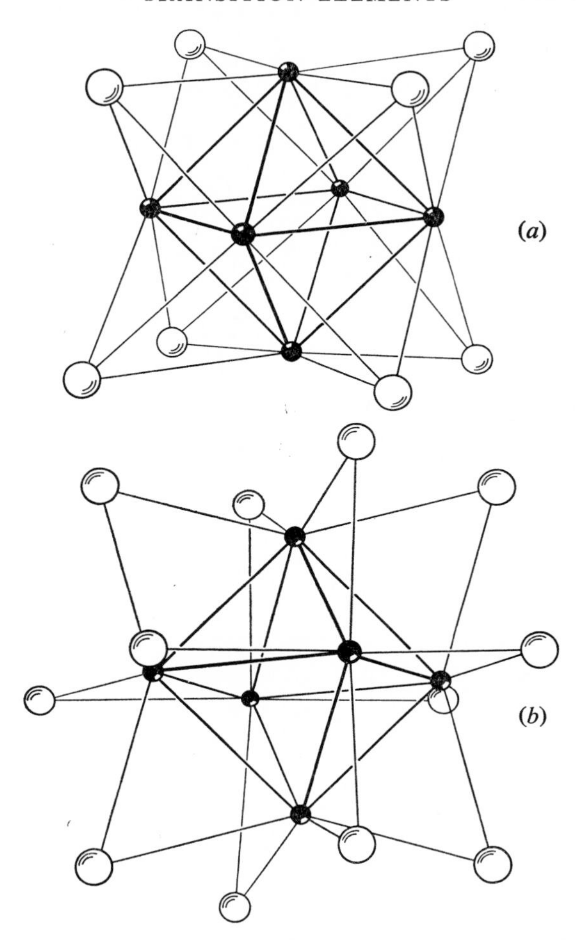

FIG. 8.40 (*a*) The structure of $Mo_6Cl_8^{4+}$; (*b*) the structure of $Ta_6Cl_{12}^{2+}$.

eight three-centre bonds in the faces of the octahedron. The twelve chlorines are bridging the edges of the octahedron, and we see that they thus complete a square antiprism arrangement of electron pairs around each tantalum (Fig. 8.41).

Another interesting type of cluster compounds are the homopoly-atomic cations such as Bi_5^{3+} and Bi_9^{5+}. Bi_9^{5+} is present in the compound that was formerly thought to be bismuth monochloride but was shown by X-ray crystallography to have the formula $Bi_{24}Cl_{28}$ and to contain the Bi_9^{5+}, $BiCl_5^{2-}$, and $Bi_2Cl_8^{2-}$ ions. The cation has a tricapped trigonal prism structure with a bismuth atom at each

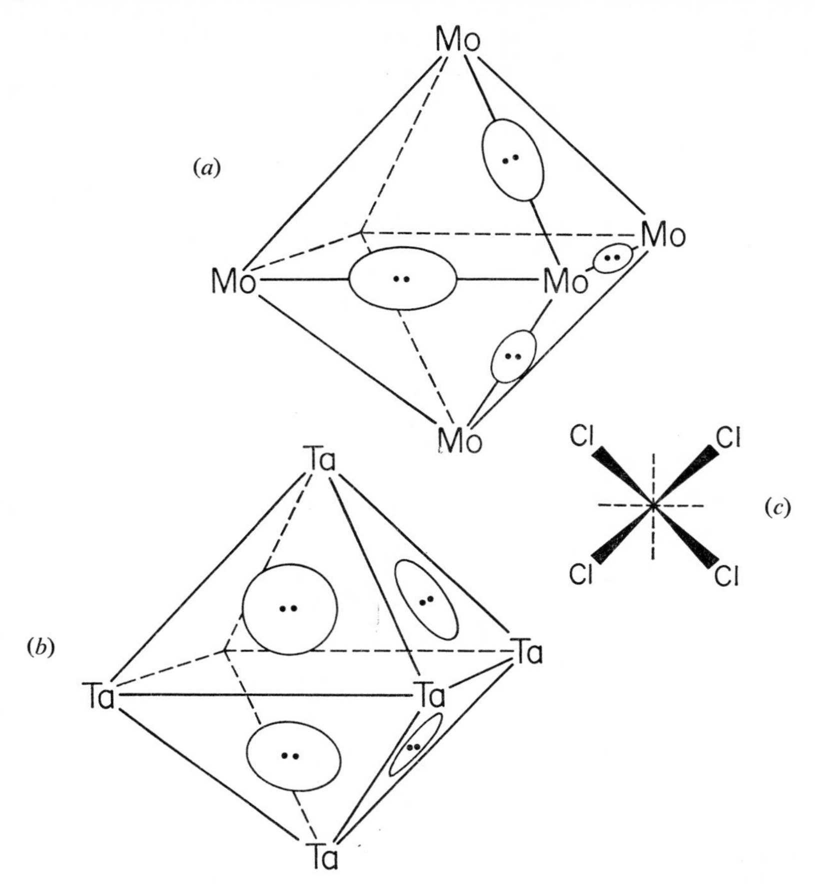

FIG. 8.41 (*a*) Two-centre metal-metal bonds in the Mo_6 cluster in Mo_6ClII^+; (*b*) three-centre metal-metal bonds in the Ta_6 cluster in $Ta_6Cl_{12}^{2+}$; (*c*) arrangement of metal-metal bonds - - - - - - and metal-chlorine bonds viewed down a fourfold axis of the octahedron in both $Mo_6Cl_8^{4+}$ and $Ta_6Cl_{12}^{2+}$.

corner of this polyhedron. As shown in Fig. 8.42 this has seven triangular faces and three approximately square faces. Bi_9^{5+} has forty electrons, or twenty electron pairs of which nine may be assigned as non-bonding pairs, one to each bismuth, leaving eleven pairs for bonding in the polyhedron. These may be assigned one to each face, giving eight three-centre and three four-centre bonds in the Bi_9^{5+} cluster .The Pb_9^{4-} cluster is isoelectronic with Bi_9^{5+} and presumably has the same structure. The structure of Bi_5^{3+} is not known but it is tempting to speculate that it is a trigonal bipyramid, as it has twenty-

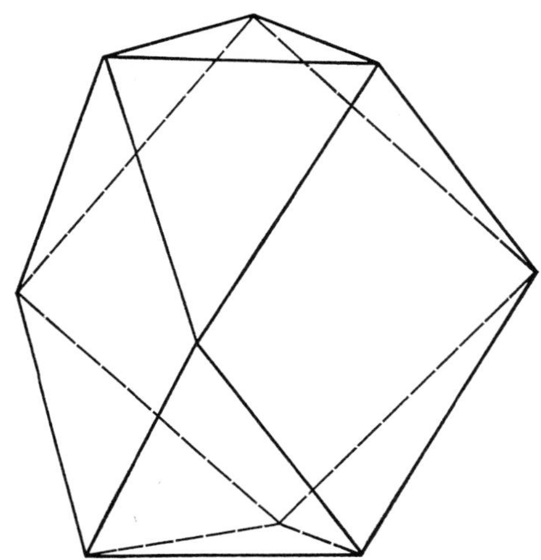

FIG. 8.42 The tricapped trigonal prism structure of Bi_9^{5+}.

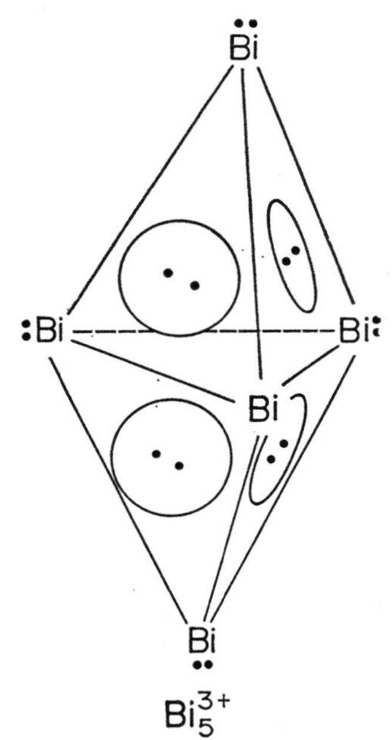

FIG. 8.43 Proposed structure and bonding of the Bi_5^{3+} ion.

two electrons or eleven pairs, five of which may be assumed to be non-bonding pairs, one to each bismuth, and the remaining pairs may be assigned one to each of the six triangular faces of the trigonal bipyramid (Fig. 8.43).

Many other cluster compounds are known particularly among the transition metal carbonyls but the detailed discussion of their structures is beyond the scope of this book. Moreover their structures cannot always be easily understood in terms of the rather simple ideas discussed above.

REFERENCES AND SUGGESTIONS FOR FURTHER READING

Tables of Interatomic Distances and Configuration in Molecules and Ions, Special Publication No. 11, The Chemical Society, London (1958).

Supplement, Special Publication No. 18, The Chemical Society, London (1965).

F. A. COTTON and G. WILKINSON, *Advanced Inorganic Chemistry*, 2nd Ed., Interscience, 1966.

E. L. MUETTERTIES and R. A. SCHUNN, *Quart. Rev. Chem. Soc.*, **20,** 245 (1966).

L. PAULING, *Nature of the Chemical Bond*, 3rd Ed., Cornell University Press, 1960.

A. F. WELLS, *Structural Inorganic Chemistry*, 3rd Ed., Oxford University Press, 1962.

9

Comparison of the Localized Electron Pair Model with Other Theories of Chemical Bonding and Molecular Structure

For the past thirty or more years it has been customary for chemists to discuss molecular geometry in terms of the directional properties of atomic orbitals or appropriate sets of hybrid orbitals derived from these atomic orbitals. This method, which was developed primarily by Pauling and is often known as the valence-bond method, has achieved wide popularity and most discussions of molecular structure have been given in these terms. More recently considerable attention has been given to the molecular orbital theory in which the electrons in a molecule are described as occupying orbitals which embrace the whole molecule. Although this theory has been very useful for the description of the energy states of a molecule it has been less successful and less popular for the discussion of molecular geometry although its use for this purpose is certainly increasing. It is important therefore to discuss the relationship of these alternative approaches to molecular geometry with that discussed in this book.

9.1 ATOMIC ORBITALS

An electron is described by a wave-function ψ, the explicit form of which for any system can be obtained by solving the Schrodinger equation for that system. The physical significance of the wave function can be described in two alternative but entirely equivalent ways. If it is assumed that the electron has no real position but is diffused over the entire space occupied by the wave then ψ^2 at a given point is proportional to the density of the electron at that point. If on the other hand it is assumed that the electron has a definite position at any instant, the Heisenberg uncertainty principle indicates that this position cannot be precisely determined, and ψ^2 at a point is then proportional to the probability of finding the electron at that point.

On solving the Schrodinger equation for the hydrogen atom one obtains the form of the wave function ψ for each of the allowed energy states of the hydrogen atom. These are the atomic orbitals of the hydrogen atom. Each orbital is uniquely defined by three quantum numbers, n, l, and m. The physical significance that may be attributed to these quantum numbers is as follows:

n is a measure of the energy of the orbital and indicates the shell that the electron is in, i.e., $n = 1$ for the K shell, $n = 2$ for the L shell, etc., and it also gives the number of nodes in the orbital which is equal to $n - 1$.

l is a measure of the orbital angular momentum of the electron and it also gives the number of planar nodes, i.e., it determines the shape of the orbital. The possible values of l are $n - 1, n - 2, \ldots, 0$. In describing any orbital the principal quantum number n is given first followed by a letter denoting the value of l according to the following code:

$$l = 0, 1, 2, 3, 4, 5$$
$$s \quad p \quad d \quad f \quad g \quad h$$

The third quantum number m gives the orientation of the orbital with respect to some fixed direction in space and can take the values $-l, -l - 1, 0, 1, \ldots, (l - 1), l$.

Thus for the K shell there is only one orbital, the 1s orbital, which has no nodes, and ψ has a maximum at the nucleus and decreases with increasing distance from the nucleus (Fig. 9.1). Such an orbital can be represented by a set of concentric spheres surrounding the nucleus, each sphere representing a surface of constant ψ. Since the distribution of electron density is given by ψ^2 a similar plot and set of spheres represents the distribution of electron density in the orbital. A cross section through the nucleus then gives a set of circular electron probability (or electron density) contours (Fig. 9.1). The shape of the orbital may be simply represented by one such contour drawn to include a large fraction of the total electron density, e.g., 90% or 99%.

For the L shell $n = 2$ and l can take the values 0 and 1. For $l = 1$ there are three possible values of $m = -1, 0,$ and $+1$. Thus there is one 2*s* orbital and three 2*p* orbitals. Since each orbital with $n = 2$ has one node there is a spherical node in the case of the 2*s* orbital but this does not change the angular distribution of the electron which can still be represented by a sphere. The *p* orbitals each have a planar node through the nucleus and they may be conveniently represented by contours of constant electron density as shown in Fig. 9.1. Each orbital has cylindrical symmetry around its axis and

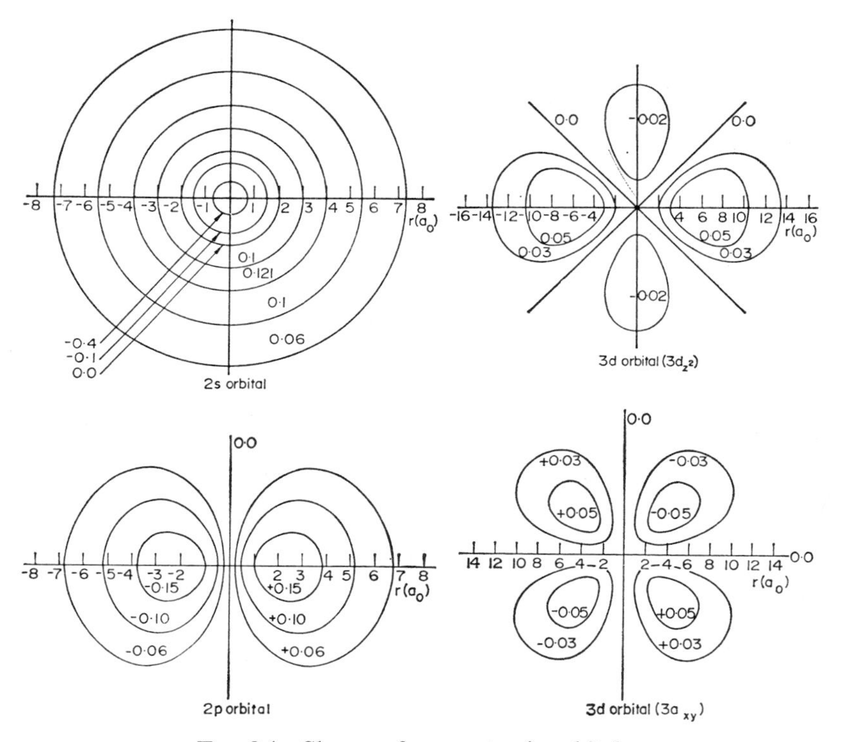

FIG. 9.1 Shapes of some atomic orbitals.

the three *p* orbitals corresponding to the three possible values of the quantum number *m* are described as the $2p_x$, $2p_y$, and $2p_z$ orbitals. For the M shell $n = 3$ and the possible values of l are 0, 1, and 2. Thus the M shell contains one 3*s* orbital, three 3*p* orbitals, and five 3*d* orbitals which have respectively zero, one, and two planar nodes. The shapes of the 3*d* orbitals are shown in Fig. 9.1.

It is assumed that other atoms have orbitals of the same general form and that electrons can be accommodated in these orbitals two at a time provided that they have opposite spin in accordance with the Pauli Exclusion principle. In atoms other than hydrogen the *s*, *p*, and *d* orbitals in the same quantum level no longer have the same energy because of shielding effects, and the order of energies is 1*s*, 2*s*, 2*p*, 3*s*, 3*p*, 4*s*, 3*d*, etc., leading to the electronic configurations given in Table 8.1 for the elements potassium to krypton.

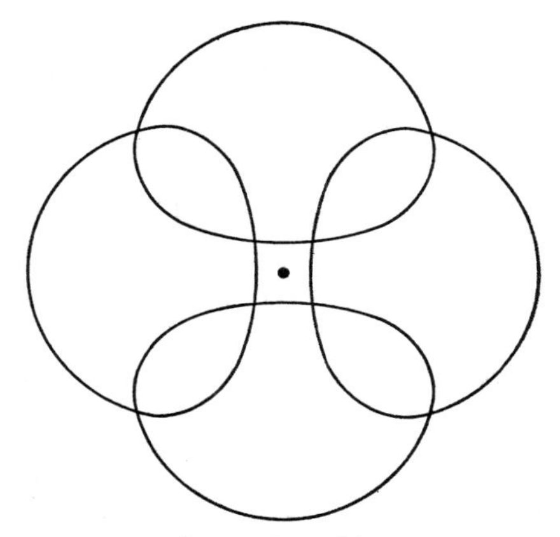

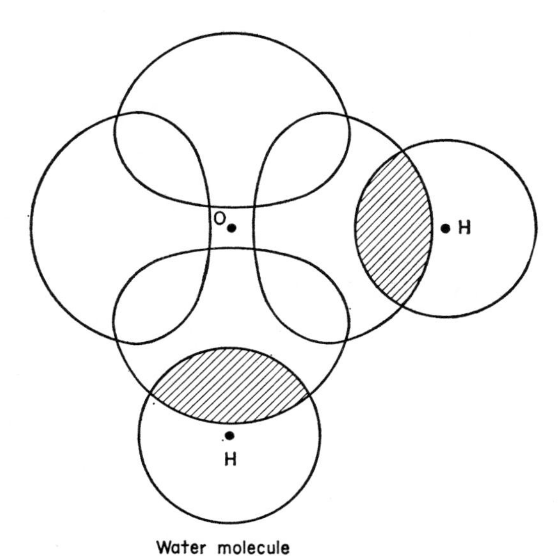

FIG. 9.2 Valence-bond description of the water molecule.

9.2 OVERLAP OF ATOMIC ORBITALS AND BOND FORMATION

According to the valence bond method for describing molecules a singly occupied orbital on one atom can overlap with a singly occupied orbital on another atom, the overlap region constituting a region of high probability of finding the two electrons. The increased electron density in this region then provides the electrostatic attraction holding the two nuclei together. Alternatively it may be imagined that the two orbitals combine, i.e., add together, to form a localized bond orbital embracing the two nuclei and containing a pair of electrons, which constitutes the covalent bond between the two nuclei.

Since the *p* orbitals have directional character the directional characteristics of chemical bonds have been associated with the *p* atomic orbitals. Thus oxygen has been supposed to use its two singly occupied *p* orbitals to overlap with, for example, two singly occupied hydrogen orbitals to form two OH bonds at right angles thus leading to an angular molecule (Fig. 9.2). Similarly the three singly occupied orbitals of the nitrogen atom may be used to form three bonds at right angles thus giving a pyramidal geometry to the three valent compounds of nitrogen.

9.3 HYBRID ORBITALS

The valence-bond method encounters two apparent difficulties when carbon is considered. First the ground state of the carbon atom is $1s^2 2s^2 2p_x 2p_y$ and therefore carbon should form only two bonds whereas carbon in general forms four bonds. It clearly must form these bonds using an excited state with four unpaired electrons. This is the $1s^2 2p_x 2p_y 2p_z$ state which lies some 97 kcal above the ground state. Second the three *p* orbitals might be expected to form three bonds at right angles leaving the spherical *s* orbital to form a bond in some unspecified direction and this is not the observed tetrahedral arrangement of four bonds. Pauling and Slater pointed out that a combination of these atomic orbitals could be formed which would be as concentrated as possible in one direction, i.e., would have maximum overlap with another orbital in this direction and therefore presumably lead to the strongest possible bond (Fig. 9.3). When this is done it is found that three further orbitals which are equivalent to the first can be constructed and these lie in the four tetrahedral directions. Alternatively one may just require that the four orbitals should all be equivalent. In any case the same set of tetra-

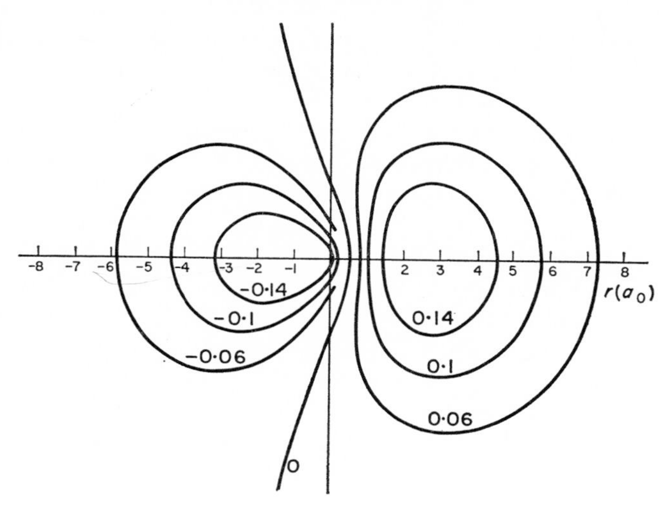

FIG. 9.3 An sp^3 hybrid orbital.

hedrally directed orbitals is obtained. They are termed sp^3 hybrid orbitals. With appropriate choice of axes these orbitals have the simple form:

$$\begin{aligned}
\psi_{sp^3}(1) &= \tfrac{1}{2}(\psi_{2s} + \psi_{2px} + \psi_{2py} + \psi_{2pz}) \\
\psi_{sp^3}(2) &= \tfrac{1}{2}(\psi_{2s} + \psi_{2px} - \psi_{2py} - \psi_{2pz}) \\
\psi_{sp^3}(3) &= \tfrac{1}{2}(\psi_{2s} - \psi_{2px} + \psi_{2py} - \psi_{2pz}) \\
\psi_{sp^3}(4) &= \tfrac{1}{2}(\psi_{2s} - \psi_{2px} - \psi_{2py} + \psi_{2pz}).
\end{aligned}$$

This set of orbitals is entirely equivalent to the set of atomic orbitals from which they are constructed, but they have the advantage that they enable the formation of tetrahedral bonds by carbon to be represented in a more satisfactory manner. Similarly, it is possible to form the hybrid orbital combinations of one s and one p orbital which have the following form.

$$\psi_{sp}(1) = \frac{1}{\sqrt{2}}(\psi_{2s} + \psi_{2x} + \psi_{2px})$$

$$\psi_{sp}(2) = \frac{1}{\sqrt{2}}(\psi_{2s} - \psi_{2x} - \psi_{2px})$$

They are called sp hybrid orbitals and are shown in Fig. 9.4. A third set of hybrid orbitals can be constructed from one s and two p orbitals, and these are directed in a plane at 120° to each other and

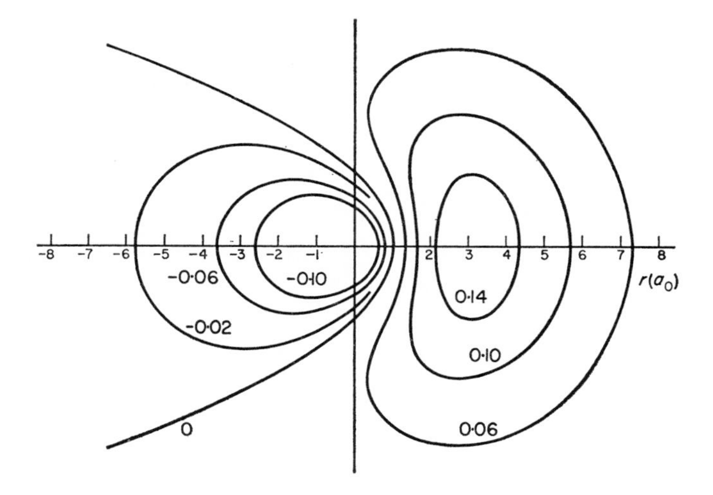

FIG. 9.4 An *sp* hybrid orbital.

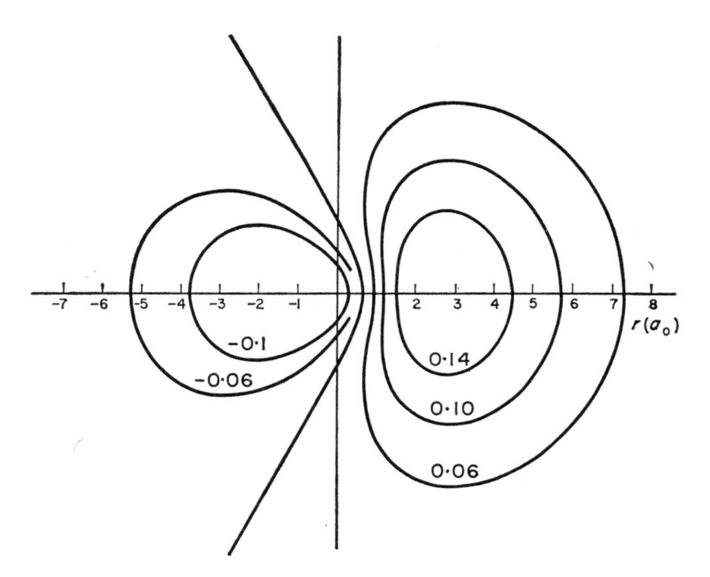

FIG. 9.5 An sp^2 hybrid orbital.

are called sp^2 hybrid orbitals (Fig. 9.5). The linear *sp* hybrid orbitals are used to describe the bonds in a linear molecule such as BeH_2 (Fig. 9.6) which is formed from the excited $2s2p$ state of the beryllium, and the sp^2 set of hybrid orbitals are used to describe the bonds in a molecule such as BH_3 which is formed from the excited $2s2p_x2p_y$ state of the boron atom (Fig. 9.6).

As the simple descriptions of the water molecule and the ammonia molecule lead to a predicted bond angle of 90° which is considerably smaller than the observed angles of 104·5° and 107·5° it is usually assumed that a set of four sp^3 hybrid orbitals also provide a better

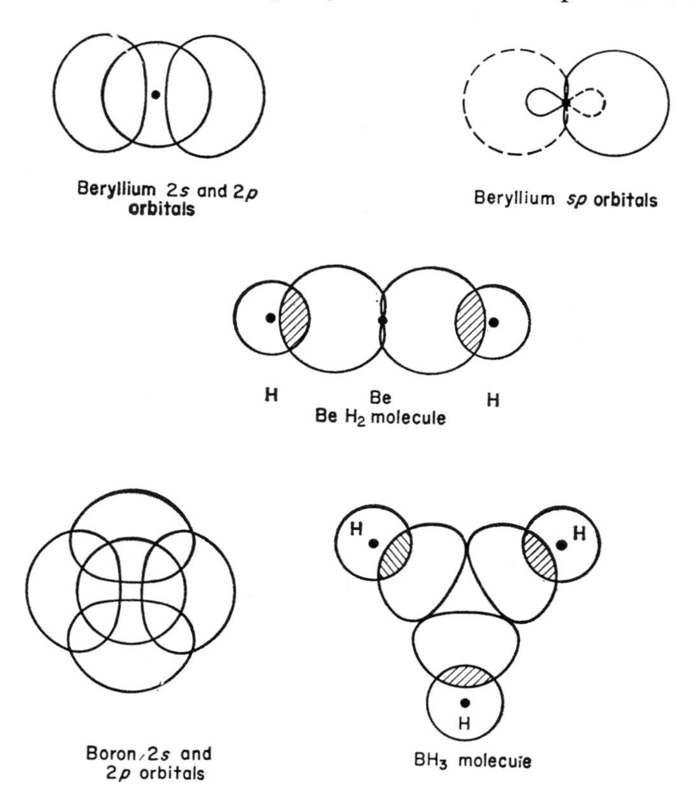

FIG. 9.6 Representation of the formation of the BeH_2 and BH_3 molecules according to the valence-bond method.

description of the bonding in these molecules as well. Thus in the water molecule two of the sp^3 hybrid orbitals are used for bonding and the other two accommodate the two non-bonding electron pairs. Similarly in ammonia three of the tetrahedral orbitals are used for bonding and one for the lone-pair. This description of these molecules is of course very similar to the description used in this book based on the tetrahedral arrangement of four localized electron pairs.

9.4 PAULI EXCLUSION PRINCIPLE AND ELECTRON DISTRIBUTION

A limitation of the atomic orbital description of a molecular system is the implicit assumption that the electrons are occupying the orbitals independently, whereas this cannot of course be the case. The electrons interact with each other because of electrostatic repulsion and, perhaps more importantly, because of the operation of the Pauli exclusion principle. This states that the wave function ψ for any system must be antisymmetric to the interchange of the co-ordinates of any two electrons. In the case that these two electrons have the same spin the spin part of the wave function is symmetric and hence the space part must be antisymmetric, i.e.,

$$\psi(x_1, x_2, x_3 \ldots) = -\psi(x_2, x_1, x_3 \ldots)$$

Now if two electrons have the same co-ordinates $x_1 = x_2 = x$ then

$$\psi(x, x, x_3 \ldots) = -\psi(x, x, x_3 \ldots)$$

and hence

$$(x, x, x_3 \ldots) = 0.$$

Thus two electrons with the same spin cannot be at the same point in space and in general they tend to avoid each other. If two electrons occupying separate atomic orbitals are assumed to be independent the total wave function can be represented as a product of the atomic wave functions. However such a simple product wave function is not a good wave function in that it does not obey the Pauli exclusion principle. Thus if electron 1 is in orbital a and electron 2 with the same spin is in orbital b the function $\psi = a(1)b(2)$ is not antisymmetric to electron interchange since $a(1)b(2) \neq -a(2)b(1)$. Allowance for the operation of the exclusion principle can however easily be made by taking appropriate antisymmetrical combinations of the atomic wave functions. For two electrons with the same spin occupying orbitals a and b the appropriate antisymmetrical wave function is

$$\psi = a(1)b(2) - a(2)b(1)$$

which may conveniently be written in the form of the determinant

$$\psi = \begin{vmatrix} a(1) & a(2) \\ b(1) & b(2) \end{vmatrix}.$$

It is clear that this function does have the property of antisymmetry because

$$a(2)b(1) - a(1)b(2) = -[a(1)b(2) - a(2)b(1)],$$

or simply because a determinant changes sign when any two columns are interchanged. Using such an antisymmetric wave function we can calculate the most probable distribution of the electrons in any system.

We will consider as an example two electrons with the same spin, one in a $2s$ orbital and the other in a $2p$ orbital which we will take to be the $2p_z$ orbital. As our interest is mainly in the angular distribution of the electrons we may for simplicity use hydrogen-like wave functions and take their radial parts to be identical. Thus we may write $\psi_{2s} = R$ and $\psi_{2p} = \sqrt{3}R \cos \theta$. The total wave function is then

$$\psi = R(1)\sqrt{3}R \cos \theta(2) - R(2)\sqrt{3}R \cos \theta(1)$$

and the angular dependence can then be written as $\psi(\theta) =$ constant$(\cos \theta(2) - \cos \theta(1))$. Hence $\psi^2(\theta) =$ constant$(\cos \theta(2) - \cos \theta(1))^2$.

This function has a maximum value when either $\theta(1) = 0°$ and $\theta(2) = 180°$ or when $\theta(1) = 180°$ and $\theta(2) = 0°$. Thus we find that the most probable relative distribution of the two electrons when allowance is made for the operation of the Pauli exclusion principle is with the two electrons at 180° from each other. The same conclusion is reached more readily if we describe the system in terms of the *sp* hybrid orbitals since these orbitals have their maxima at 180° from each other. Thus the advantage of the *sp* hybrid orbital description of the system is that it gives a more obvious picture of the relative distribution of the two electrons than does the atomic orbital description. This is because the two atomic orbitals overlap extensively in space and the simple product wave function is accordingly a very poor wave function for the system (Fig. 9.6). Hence the true relative electron distribution cannot be obtained by considering the electrons to occupy the $2s$ and the $2p$ orbitals independently. The extensive region of overlap of the two orbitals is a region of space where, if the two electrons moved independently, they would have a finite probability of being found at the same point in space, and this contravenes the Pauli exclusion principle. On the other hand the simple product function of the two hybrid orbitals is a relatively good wave function because the two orbitals overlap each other to a small extent and accordingly the distribution of one electron may be regarded as being very largely independent of the other.

For the atomic orbital description ψ^2 the probability that the two electrons will be found simultaneously in any given positions is given by

$$\begin{aligned}\psi^2 &= (s(1)p(2) - s(2)p(1))^2 \\ &= (s(1)^2p(2)^2 + s(2)^2(1)^2) - 2s(1)p(2)s(2)p(1) \\ &= P_c - P_e.\end{aligned}$$

The term P_c may be regarded as corresponding to a classical interpretation of an electron in each of the distributions s^2 and p^2 (allowing for either electron to be in either orbital). The term P_e corresponds to the non-classical or 'exchange' contribution to the probability distribution. Its magnitude, which is proportional to the overlap between the two orbitals, determines the amount by which the true distribution differs from the 'classical' distribution. If the two electrons are in the same region of overlap of the two orbitals then they are relatively close together; in this case P_e is large and positive and accordingly $\psi^2 = P_c - P_e$ is small, i.e. the probability of finding the two electrons in the same region of overlap of the two orbitals is small. On the other hand, if the two electrons are in different regions of overlap then P_e is negative, because the p orbital has opposite sign in the two regions and the probability ψ^2 is large. Again we see the tendency for the two electrons to keep apart. Now for the hybrid orbital description the probability function is given by

$$\begin{aligned}\psi^2 &= (sp_1(1)sp_2(2) - sp_1(2)sp_2(1))^2 \\ &= (sp_1(1)^2sp_2(2)^2 - sp_1(2)^2sp_2(1)^2) - 2sp_1(1)sp_2(2)sp_1(2)sp_2(1) \\ &= P_c' - P_e'.\end{aligned}$$

The total probability function ψ^2 is of course identical with that calculated from atomic orbitals, but the relative contribution of P_c' and P_e' are not the same as those of P_c and P_e. For the hybrid orbital description the first term P_c' may again be regarded as corresponding to a classical interpretation of the distribution of the electrons, with one in each orbital, and the second term P_e' is the non-classical or 'exchange' contribution which determines the amount by which the true distribution differs from the classical distribution. Again the 'exchange' term depends on the overlap of the two orbitals, and as this is small P_e is small and hence the classical distribution corresponds reasonably closely to the true distribution.

In general a classical interpretation of the distribution of electrons in hybrid orbitals corresponds more closely to the most probable distribution than does a similar classical interpretation of the distribution of the electrons in atomic orbitals. In the limit, a set of completely localized orbitals which did not overlap would correspond exactly to the electron distribution in the system because the exclusion principle would have been automatically allowed for.

9.5 SOME DIFFICULTIES ASSOCIATED WITH HYBRID ORBITALS

The hybrid orbital method can be extended to describe the geometry associated with higher co-ordination numbers such as five and six by the inclusion of *d* orbitals. Thus it can be shown that six octahedrally directed hybrid orbitals can be constructed from the $sp^3d_{x^2-y^2}d_{z^2}$ set of atomic orbitals, a trigonal bipyramidal set of five hybrid orbitals from the $sp^3d_{z^2}$ set of atomic orbitals and a square pyramidal set of five hybrid orbitals from the $spd_{x^2-y^2}$ set of atomic orbitals. Some further difficulties associated with the hybrid orbital method now become apparent. Firstly, the geometry associated with a set of hybrid orbitals depends on the particular *d* orbital chosen and there is no *a priori* way of making this choice in any given case. Secondly, in a case such as the trigonal bipyramid set of hybrids which are not all equivalent as the three equatorial orbitals cannot be equivalent to the two axial orbitals it is not possible to give an explicit form to these orbitals as the relative contributions of the *s* and d_{z^2} orbitals to the two axial or three equatorial orbitals can be varied in an arbitrary manner.

The overlap of a hybrid orbital with a ligand orbital is only a very approximate description of a bonding orbital and it can be shown that orbitals formed by such a linear combination are not satisfactory solutions of the Schrodinger equation as they are not orthogonal. Thus the method of hybrid orbitals is simply a method of constructing from the atomic orbitals of an atom a set of localized orbitals that take account of the Pauli principle and that correspond approximately to the chemist's conception of independent chemical bonds and localized lone-pairs that have a definite spatial relationship to each other. It would seem therefore to be at least as reasonable to base a description of electrons in molecules directly on the Pauli exclusion principle and to make use of the fact that electrons of the same spin tend to avoid each other and thus to occupy separate regions of space. The idea that an electron of a given spin is surrounded by a region of space from which it excludes other electrons of the same spin has long been recognized and has been called a Fermi hole. Indeed this idea plays an important role in the procedure used by Hermann and Skilma for the calculation of the charge distribution in atoms by the self-consistent field method. This region of space may as a first approximation be taken to be spherical and can thus be identified with the hard sphere orbitals that have been used in this book. In order to satisfy the operation of the exclusion principle these Fermi holes or hard-sphere orbitals must then be arranged so that they do not overlap with each other and indeed so that they keep as far apart as possible, and this of course is the basis of the electron pair arrangements that have been the basis of the discussion in this book.

9.6 MOLECULAR ORBITAL THEORY

In describing a molecule by the molecular orbital method electrons are placed in orbitals that embrace the whole molecule, i.e., in molecular orbitals. The approximate shapes and energies of these molecular orbitals are generally obtained by forming linear combinations of the appropriate atomic orbitals (LCAO method). Thus two molecular orbitals may be formed for the H_2 molecule from the $1s$ orbitals of the hydrogen atom (Fig. 9.7).

The two orbitals are described as a σ bonding orbital and a σ^* antibonding orbital. Since the electron density in this latter orbital is

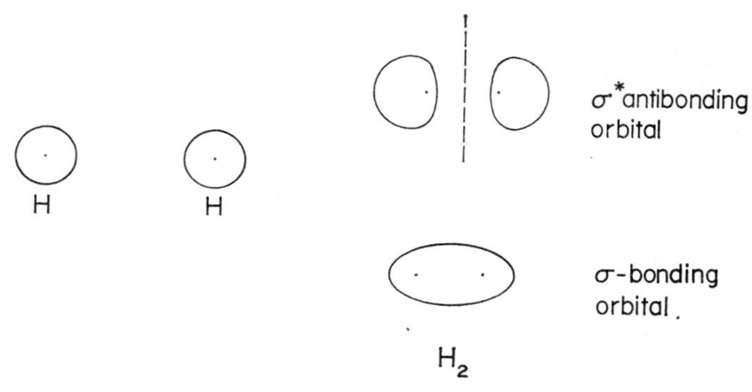

FIG. 9.7 The bonding and antibonding orbitals for the hydrogen molecule.

largely situated outside the two nuclei rather than between them it tends to pull the two nuclei apart and is therefore antibonding. In H_2 the antibonding orbital is empty and consequently a stable molecule results. The same orbitals can be used to describe He_2, and in this case as there are four electrons the antibonding orbital must also be occupied; consequently there is no resultant bonding between the two nuclei, i.e., He_2 is not a stable molecule. Combinations may also be made of atomic p orbitals to give both σ- and π-type molecular orbitals (Fig. 9.8). Hence for a number of simple diatomic molecules the energy level scheme given in Fig. 9.9 may be used. Thus nitrogen which has ten valency electrons has the following configuration:

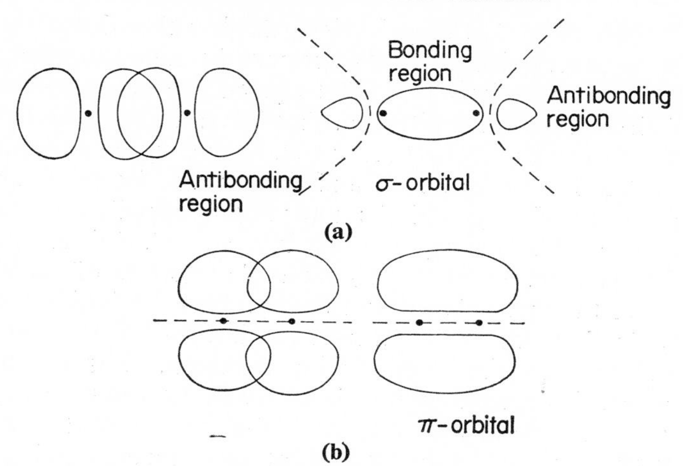

FIG. 9.8 Formation of σ and π orbitals by the overlap of atomic p orbitals: (*a*) formation of a σ orbital by end-on overlap of two p-orbitals; (*b*) formation of a π orbital by sideways overlap of two p-orbitals.

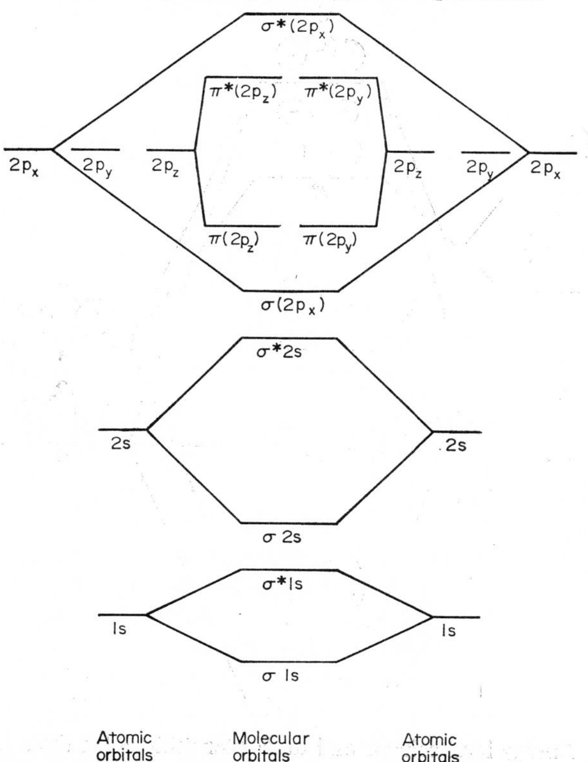

FIG. 9.9 Simple qualitative energy level scheme for a diatomic molecule, e.g., N_2.

$(\sigma 2s)^2(\sigma^* 2s)^2(\sigma 2p)^2(\pi 2p)^2(\pi 2p)^2$. Since the bonding due to the $\sigma 2s$ electrons is cancelled by the two electrons in the antibonding $\sigma^* 2s$ orbital these two orbitals are in fact equivalent to two non-bonding pairs of electrons on each nitrogen. This leaves a total of three bonding pairs of electrons, i.e., a triple bond. In the oxygen molecule two more electrons must be added and these must clearly occupy the degenerate pair of $\pi^* 2p$ orbitals; according to Hund's rule they will occupy these orbitals singly and will be unpaired. This explanation of the paramagnetism of the O_2 molecule was one of the earliest successes of the molecular-orbital theory. Thus, excluding the non-bonding $2s$ electrons the oxygen molecule contains three bonding pairs and two unpaired antibonding electrons. It is exactly this situation that is depicted by the electron-dot diagrams on page 86.

The general form of the molecular orbitals for polyatomic molecules may be obtained by the methods of group theory making use of

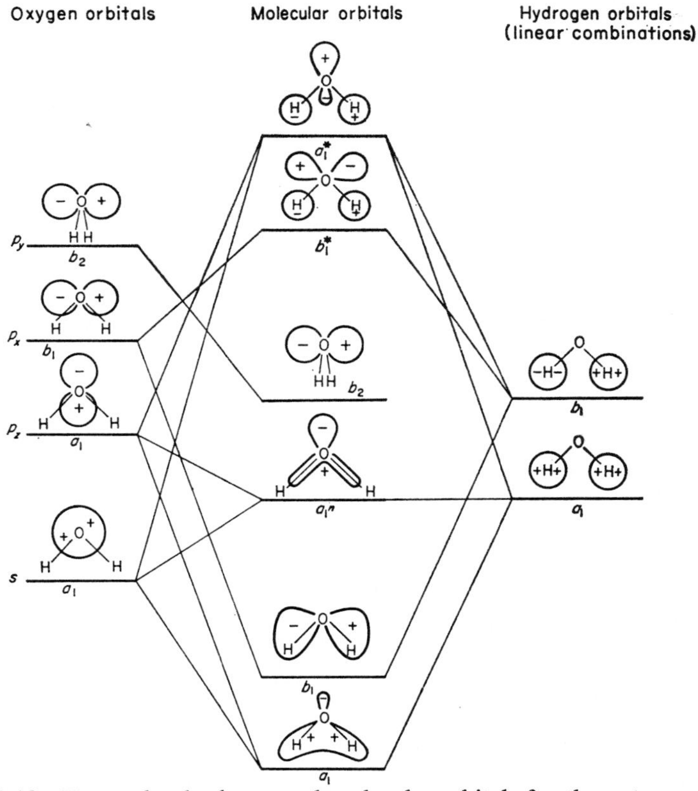

FIG. 9.10 Energy level scheme and molecular orbitals for the water molecule.

the symmetry properties of the molecule. The molecular orbitals for the water molecule are shown in Fig. 9.10. This energy level scheme assumes that the molecule is bent. For a linear HOH molecule the energy level diagram would be as shown in Fig. 9.11. A decision as to whether the water molecule is expected to be linear or bent must then be based on a decision as to which energy level diagram will give the lowest energy for the molecule. The π orbitals in the linear

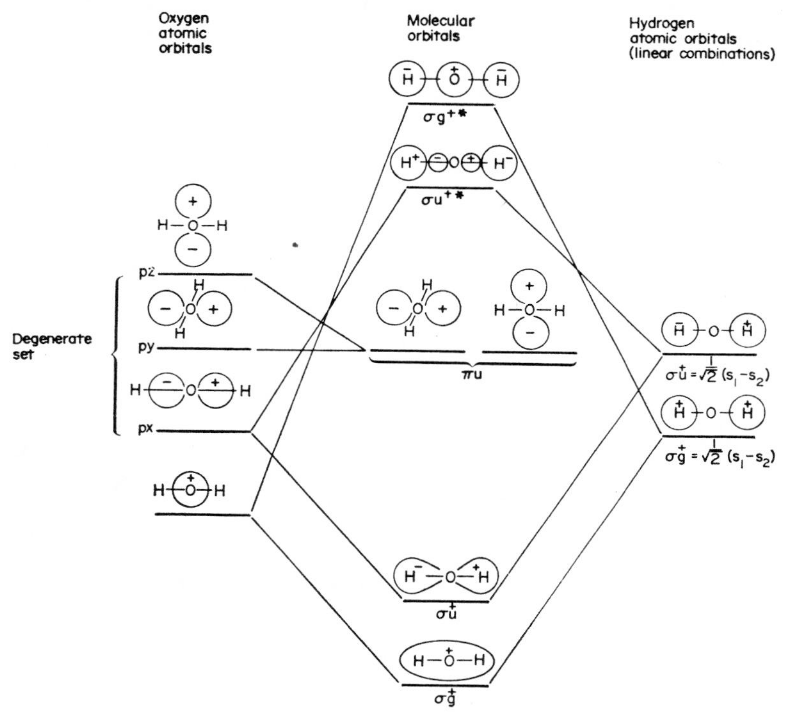

FIG. 9.11 Energy level scheme and molecular orbitals for a hypothetical linear water molecule.

molecule are in fact simply non-bonding p orbitals on oxygen, whereas in the bent molecule the b_2 orbital corresponds simply to an oxygen p orbital; but the a_1'' orbital, although largely concentrated on oxygen and therefore largely non-bonding, is in fact spread out over the whole molecule to some extent and therefore has some bonding character. Consequently one concludes that the molecule will be more stable in the bent form than in the linear form. It is clear however that this conclusion is not so easily reached as by the method discussed in this book or even by the valence bond method. In general for any polyatomic molecule the prediction of shape can only be made by considering the relative energies of different plausible shapes. For complex molecules, particularly if they are of low symmetry, the estimation of the relative energies of the various molecular orbitals is a matter of some difficulty and uncertainty and the prediction of molecular shape is correspondingly uncertain. It is probably not unfair to say that the molecular orbital method is not yet a generally useful theory for the general prediction of molecular shapes.

We note that in the water molecule the linear form minimizes the interaction between the bonding electron pairs but concentrates the two non-bonding pairs of electrons into the oxygen *p* orbitals, thereby maximizing their interaction. When the molecule bends the two non-bonding electron pairs can move apart somewhat thus decreasing their interaction. In the other limit represented by the simple valence bond theory the two bonds are formed by two *p* orbitals leaving the lone-pair electrons in a 2*s* and a 2*p* orbital on the oxygen, i.e., at a maximum distance apart, as may be seen by using the alternative but equivalent *sp* hybrid orbitals. The actual approximately tetrahedral bond angle results from the minimizing of the interactions between all four pairs of electrons in the valency shell of oxygen which may to a reasonable approximation be regarded as four essentially equivalent pairs. This is the idea that forms the basis of this book.

REFERENCES AND SUGGESTIONS FOR FURTHER READING

R. DAUDEL, *The Fundamentals of Theoretical Chemistry*, Pergamon Press, 1968.

R. J. GILLESPIE and R. S. NYHOLM, *Progress in Stereochemistry*, Vol. II, Butterworth, London, 1958.

J. E. LENNARD-JONES, *Adv. Sci.* **51,** 136, 1954.

J. W. LINNETT, *Wave Mechanics and Valency*, Methuen, 1960.

W. KAUZMANN, *Quantum Chemistry*, Academic Press, 1957.

Index

159